ISSUES IN ENVIRONMENTAL SCIENCE
AND TECHNOLOGY

EDITORS: R. E. HESTER AND R. M. HARRISON

2

Waste Incineration and the Environment

ROYAL
SOCIETY OF
CHEMISTRY

ISBN 0-85404-205-9
ISSN 1350-7583

A catalogue record for this book is available from the British Library

© The Royal Society of Chemistry 1994

Published by the Royal Society of Chemistry, Thomas Graham House,
Science Park, Cambridge CB4 4WF

Typeset by Vision Typesetting, Manchester
Printed and bound in Great Britain by Bath Press, Bath

Preface

As one of us has been involved recently as an expert witness in a Public Enquiry into a proposed incinerator development, we are acutely aware of the passions which are aroused in the local population by the proposed development of an incinerator. Yet, with modern pollution control devices, these combustion plants can be very clean indeed, and vastly cleaner than the much smaller municipal incinerators which were built in the UK in the 1960s. Modern incinerators also incorporate heat recovery ('energy from waste') and hence come quite high on the list of desirable options for waste disposal following recycling, which everyone sees as desirable, but the practicability of which has yet to be established in depth. In this Issue we seek to explore the options for waste disposal, the emissions and consequences of incineration, and approaches to regulation.

The Issue starts with an overview of waste management options by Judith Petts which examines the advantages and disadvantages of different options and places incineration in its context as a popular and economic means of disposing of wastes. This is followed by a chapter by Paul Williams, reviewing in detail the pollutant emissions which arise from incineration (mainly atmospheric, but also giving consideration to water and solid residues), the mechanisms by which such pollutants are generated, and the amounts typically emitted. A case study by George Rae of the proposed Cory development at Belvedere, East London, follows, which shows us exactly what emissions are to be expected from a modern waste-to-energy plant and the engineering controls which are used to limit them.

The next two articles deal with specific types of emissions from waste incineration. Gev Eduljee addresses the organic micropollutants, their chemical nature, formation mechanisms, and consequences, and Gregory Carroll pilot-scale research on trace metal emissions. Such research is valuable in optimizing pollution control systems. These chapters are followed by a contribution by E. Malone Steverson describing the United States regulatory approach to incineration emissions. The development of this approach is described and offers considerable insights to those unfamiliar with it. The final chapter by Owen Harrop examines means for assessment of the environmental impact of pollutants emitted from incineration and gives a valuable introduction to EIA methodology as applied to incinerator developments.

We believe that this compilation of papers provides a comprehensive overview of the current state of knowledge with regard to incineration of wastes which should prove of value to scientists, legislators, consultants, and industries with waste disposal problems. We are grateful to the authors for producing excellent contributions on a very short timescale.

Ronald E. Hester
Roy M. Harrison

Contents

Contents

Editors

Ronald E. Hester, BSc, DSc(London), PhD(Cornell), FRSC, CChem

Ronald E. Hester is Professor of Chemistry in the University of York. He was for short periods a research fellow in Cambridge and an assistant professor at Cornell before being appointed to a lectureship in chemistry in York in 1965. He has been a full professor in York since 1983. His more than 200 publications are mainly in the area of vibrational spectroscopy, latterly focusing on time-resolved studies of photoreaction intermediates and on biomolecular systems in solution. He is active in environmental chemistry and is a founder member and former chairman of the Environment Group of the Royal Society of Chemistry and editor of 'Industry and the Environment in Perspective' (RSC, 1983) and 'Understanding Our Environment' (RSC, 1986). As a member of the Council of the UK Science and Engineering Research Council and several of its sub-committees, panels, and boards, he has been heavily involved in national science policy and administration. He was, from 1991–93, a member of the UK Department of the Environment Advisory Committee on Hazardous Substances and is currently a member of the Scientific Affairs Board of the Royal Society of Chemistry.

Roy M. Harrison, BSc, PhD, DSc (Birmingham), FRSC, CChem, FRMetS, FRSH

Roy M. Harrison is Queen Elizabeth II Birmingham Centenary Professor of Environmental Health in the University of Birmingham. He was previously Lecturer in Environmental Sciences at the University of Lancaster and Reader and Director of the Institute of Aerosol Science at the University of Essex. His more than 200 publications are mainly in the field of environmental chemistry, although his current work includes studies of human health impacts of atmospheric pollutants as well as research into the chemistry of pollution phenomena. He is a former member and past Chairman of the Environment Group of the Royal Society of Chemistry for whom he has edited 'Pollution: Causes, Effects and Control', (RSC, 1983; Second Edition, 1990) and 'Understanding our Environment: An Introduction to Environmental Chemistry and Pollution' (RSC, Second Edition, 1992). He has a close interest in scientific and policy aspects of air pollution, currently being Chairman of the Department of Environment Quality of Urban Air Review Group as well as a member of the DoE Expert Panel on Air Quality Standards and Photochemical Oxidants Review Group and the Department of Health Committee on the Medical Effects of Air Pollutants.

Contributors

G. Carroll, *US Environmental Protection Agency, Office of Research and Development, 26 W. Martin Luther King Drive, Cincinnati, Ohio 45268, USA*

G. H. Eduljee, *Environmental Resources Management, Eaton House, Wallbrook Court, North Hinksey Lane, Oxford OX2 0QS, UK*

D. O. Harrop, *Aspinwall and Company, Walford Manor, Baschurch, Shropshire SY4 2HH, UK*

J. Petts, *Centre for Extension Studies, Loughborough University of Technology, Loughborough, Leicestershire LE11 3TU, UK*

G. Rae, *Cory Environmental Limited, 25 Wellington Street, London WC2E 7DA, UK*

E. M. Steverson, *Science Applications Int., Corp., 545 Shoup Avenue, PO Box 50697, Idaho Falls, Idaho 83405-0697, USA*

P. Williams, *Department of Fuel and Energy, University of Leeds, Leeds LS2 9JT, UK*

Incineration as a Waste Management Option

J. PETTS

1 Introduction

Role of Incineration

For thousands of years the value of burning wastes has been recognized, both to reduce the quantity of surplus materials generated by households, trades, and agricultural practices, and to provide fuel for heating or cooking. Recognition of the potential environmental problems generated by burning wastes also has a long history. In the United Kingdom (UK) the existence of city controls on the burning of rubbish in open dumps can be traced back to the 13th century.

The industrial revolution and accompanying urban population explosion of the 18th and 19th centuries transformed the nature and volume of wastes arisings and the potential health problems of improper disposal practices. Mass-burning of wastes in enclosed and controlled conditions became an important waste management option. The first municipal solid waste (MSW) incinerator in England was commissioned at Nottingham in 1874,[1] and by 1912 there were some 300 incinerators in the UK, 76 generating power from waste.[2] Similar early developments took place in other countries, including Sweden, Germany, and the USA.[3-5] Provision of industrial and hazardous waste incineration capacity was primarily by the major chemical companies requiring in-house facilities; for example, in the US, Dow Chemical installed the first rotary kiln in 1948.[5] The development of large-scale commercial, or merchant, sector hazardous waste incineration capacity has primarily been post-1960s.

The specific benefits of incineration include:

(i) *A reduction in the volume and weight of waste* especially of bulky solids with a high combustible content. Reduction achieved can be up to 90% of volume and 75% of weight of materials going to final landfill.

(ii) *Destruction of some wastes and detoxification of others* to render them more suitable for final disposal, e.g. combustible carcinogens, pathologically

[1] R. Hering and S. A. Greelly, 'Collection and Disposal of Municipal Refuse', McGraw Hill, New York, 1921.
[2] A. Van Santen, *Wastes Manage. Proc.*, July 1993, 18.
[3] S. Modig, *Environ. Impact Assess. Rev.*, 1989, **9**, 247.
[4] L. Barniske, *Environ. Impact Assess. Rev.*, 1989, **9**, 279.
[5] N. Behmanesh, D. T. Allen, and J. L. Warren, *J. Air Waste Manage. Assoc.*, 1992, **42(4)**, 437.

contaminated materials, toxic organic compounds, or biologically active materials that could affect sewage treatment works.

(iii) *Destruction of the organic component of biodegradable waste* which when landfilled directly generates landfill gas (LFG). Estimates suggest that LFG may account for over 40% of the UK's total methane emissions to atmosphere.[6]

(iv) *The recovery of energy from organic wastes* with sufficient calorific value.

(v) *Replacement of fossil-fuel for energy generation* with consequent beneficial impacts in terms of the 'greenhouse' effect.

The range of wastes incinerated has expanded in many industrialized countries accompanied by development of specialized and dedicated facilities, including mobile plant.[7] Incineration development has been influenced by: (i) concerns over direct landfill of certain materials, *e.g.* clinical wastes, (ii) legislative controls curtailing other disposal routes, *e.g.* for sewage sludge, (iii) identification of new environmental problems requiring remediation, *e.g.* contaminated soils, (iv) identification of problem wastes for which incineration represents the only commercially available method of disposal, *e.g.* polychlorinated biphenyls (PCBs), and (v) recognition of energy generation potential from wastes having the potential for adverse environmental impact if inappropriately disposed, *e.g.* scrap tyres. The extent of uptake of incineration in different countries has been influenced by the availability of other disposal options, in particular landfill, and the degree of central government market intervention in, and financial support of, capital investment and operation costs.

Issues and Concerns

Despite the versatility of incineration as a waste treatment method, opposition, particularly to commercial or merchant sector facilities, has developed to such an extent over the last two decades that in many countries proposals for new plant have faced long delays and often refusal, existing plant have closed, and even national waste management programmes have had to be delayed or modified following protest (for example, that of Spain and also Australia's proposals for handling hazardous wastes). The 1970s saw a rapid growth in the concern over incineration as a public health risk, particularly with the identification of chlorinated dibenzo-*p*-dioxins (PCDDs) and dibenzofurans (PCDFs) in MSW incinerator emissions,[8] which coincided with the release of 2,3,7,8-TCDD (tetrachlorodibenzo-*p*-dioxin) and subsequent environmental contamination in a chemical accident at Seveso in Italy.

Public perceptions of health risks are underpinned by the reaction of specific communities against existing and proposed facilities in their local area, including concerns about management and control capabilities, and the management of wastes generally.[9] The balance of arguments for and against incineration forms

[6] Anon., *ENDS Report*, 1993, **217**, 7.
[7] H. E. Hesketh, F. L. Cross, and J. L. Tessitore, 'Incineration for Site Clean-Up and Destruction of Hazardous Wastes', Technomic Publishing, Lancaster, Pennsylvania, USA, 1990.
[8] K. Olie, P. L. Vermeulen, and O. Hutzinger, *Chemosphere*, 1977, **8**, 455.
[9] J. Petts, *Waste Manage. Res.*, 1992, **10**, 169.

2

the basis of national policy development. A justification of incineration as the Best Practicable Environmental Option (BPEO)[10,11] for managing different waste streams has to be set in the context of reducing the pollution potential of wastes generated and achieving the BPEO by identifying the optimum balance in terms of emissions and discharges at a reasonable cost. In 1993, the UK's Royal Commission on Environmental Pollution (RCEP) published a report on the incineration of wastes,[12] which urged the UK Government to give a higher priority to developing a national waste management strategy based upon the BPEO and commending incineration as having a more important place within such a strategy.

In order to place discussion of the role of waste incineration as the BPEO in some context it is appropriate to first consider the 'current' (*i.e.* 1993) situation with regard to its use. The potential for incineration is then considered in terms of: (i) policy development; (ii) the economics of incineration; (iii) environmental impact and risk assessment; (iv) technology development; and (v) public acceptance.

2 Use of Incineration

The UK

Table 1 presents approximate annual UK arisings for each of the main incinerable waste streams. As in many countries, the data are estimates being based upon variable arisings data collected by the local authorities at the disposal point rather than at source and in a variety of recorded formats. Figures for arisings of 'special' wastes as defined are subject to annual fluctuations, apparently partly caused by isolated disposals of contaminated soils and similar materials, but have not witnessed a significant growth compared to MSW. The UK imports hazardous wastes for treatment. Figures for the period 1991/2 show that approximately 47 000 tonnes were imported into England and Wales, of which 31% were incinerated.[13] The UK's importance in this context has been influenced by the availability of capacity for the handling of PCBs.

In 1991 there were some 230 licences for incineration facilities,[13] 47 of these held by the public sector (*i.e.* primarily the local authorities) the rest by the private sector. Less than 30 MSW incinerators were operating in 1993, five of which were recovering energy. Incineration provides for only 7% of MSW arisings. Installed capacity has shown a recent decline with existing plants not able to meet new emission standards set by Her Majesty's Inspectorate of Pollution (HMIP) under Part 1 of the Environmental Protection Act, 1990.[14]

[10] Royal Commission on Environmental Pollution, 'Air Pollution Control: An Integrated Approach', Fifth Report, Cmnd. 6371, HMSO, London, 1976.

[11] Royal Commission on Environmental Pollution, 'Best Practicable Environmental Option', Twelfth Report, Cmnd. 310, HMSO, London, 1988.

[12] Royal Commission on Environmental Pollution, 'Incineration of Waste', Seventeenth Report, Cmnd. 2181, HMSO, London, 1993.

[13] Department of the Environment, 'Digest of Environmental Protection and Water Statistics', No. 15, HMSO, London, 1993, p. 77.

[14] Her Majesty's Inspectorate of Pollution, 'Chief Inspector's Guidance to Inspectors (Environmental Protection Act, 1990): Waste Disposal and Recycling: Process Guidance Notes IPR5/3 Municipal Waste Incineration', HMSO, London, 1993.

Table 1 Estimated annual arisings of incinerable wastes in the UK[1,2]

Incinerable Wastes	Annual Arisings/ Million Tonnes
Agriculture	250[3]
—Poultry litter	1
—Straw waste	13
—Carcasses	0.1[4]
Sewage sludge (dry weight)	1[5]
Household	20
Commercial	15
Industrial	50[6]
'Special' wastes	2.7[7]
Hazardous wastes	4[8]
Clinical	0.4
Scrap tyres	0.4

[1] Department of the Environment, 'Digest of Environmental Protection and Water Statistics', No. 15, HMSO, London, 1993, p. 77.

[2] KPMG Peat Marwick McClintock, 'The Recycling and Disposal of Tyres', KPMG, London, 1990, p. 20.

[3] This figure includes over 100 million tonnes of excreta on fields.

[4] Excluding poultry carcasses.

[5] Equivalent to about 30–40 million tonnes of wet sludge.

[6] Excluding power station ash and blast furnace and steel slag.

[7] Defined in the UK by Section 17 of the Control of Pollution Act (superceded by Section 62 of the Environmental Protection Act, 1990) in terms of danger to life. To be amended to meet the requirements of EC Directive 91/689/EEC.

[8] There is no legal definition in the UK but the term is used informally to refer to 'special' wastes plus other wastes regarded as difficult to handle (primarily from industrial sources).

The plants most likely to survive are those able to recover energy and new proposals reflect recognition of the changing economics of incineration (see Section 4). MSW incinerators are generally based on agitating grates with excess air.

In 1993 there were four merchant sector chemical waste incinerators operating with a notional installed capacity of 138 000 tonnes of which 80 000 tonnes were available through rotary kiln systems. In addition, two or three in-house chemical company incinerators could accept third party wastes. In 1992, permission was granted for another 30 000 tonnes hazardous waste incinerator on Teesside, north-east England (not yet built). Approximately 3% of total 'special' wastes arisings are incinerated by the merchant sector. The majority of some 60 small chemical waste facilities within industrial companies provide for specific in-house process streams mostly utilizing liquid injection systems.

Sewage sludge has traditionally been disposed to the North Sea, used for land-spreading, or landfilled. With the ban on the former route to become effective by 1998 (under the EC Urban Waste Water Treatment Directive 91/271/EEC), incineration has become an attractive alternative option and the amount of sludge incinerated has already risen from about 45 000 tonnes in 1980 to 77 000 tonnes in 1991. A recent survey estimates that incineration could rise from the current 7% of arisings to about 19% by 2006.[15] New plant are utilizing fluidized bed systems.

le 2 Incineration in
fferent countries[1-3]

Country	% Municipal waste incinerated	No. of plant	% Incinerated municipal waste including energy recovery[4]	% Sewage sludge incinerated
Canada	9	17	7	n/a
USA	16	168	n/a	n/a
Japan	75	1900	*	n/a
Sweden	55	23	86	0
Denmark	65	38	*	19
France	42	170	67	20
Netherlands	40	12	72	10
Germany	35	47	n/a	10
Italy	18	94	21	11
Spain	6	22	61	n/a
UK	7	30	33	7

[1] Royal Commission on Environmental Pollution, 'Incineration of Waste', Seventeenth Report, Cmnd. 2181, HMSO, London, 1993, p. 23.
[2] Organization for Economic Cooperation and Development, 'Environmental Data Compendium', OECD, Paris, 1991, p. 145.
[3] The data refer to different years, mostly post 1990, although the Canadian figure relates to 1985.
[4] Includes recovery for both in-house and external use.
* References suggest most plant recover energy.

In addition to the licensed facilities, some 700–800 small incinerators for clinical waste may be operating, mostly within hospitals. As with MSW incinerators, the majority of these plant cannot meet new emission standards and are having to close. Clinical waste incineration is proving a major attraction to the private sector using ashing rotary kiln systems.

A number of small plants offer specialized services, including recent investments in plant for handling poultry litter (in Suffolk) and scrap tyres (in Wolverhampton, Midlands). Only 5% of scrap tyres are incinerated in the UK,[16] although the new plant at Wolverhampton could handle 25% of total UK arisings. Incineration of wastes at sea has now ceased in response to resolutions of the London Dumping Convention meeting of November 1990.

Other Countries

Table 2 presents comparative data on the use of incineration in other countries. The data are from a number of different sources which may have used different definitions so they should be interpreted as indicative rather than actual. Nevertheless, some revealing differences are apparent, most particularly in the

[15] Department of the Environment, 'UK Sewage Sludge Survey—1993', HMSO, London, 1993.
[16] KPMG Peat Marwick McClintock, 'The Recycling and Disposal of Tyres', KPMG, London, 1990, p. 20.

percentages of MSW incinerated, and in this context the very low usage in the UK.

The countries with over 50% of MSW incinerated reflect a shortage of landfill capacity (particularly Japan) and demands for cheap energy generation for district heating (particularly Sweden). They also reflect countries with more structured and centralized waste management planning. In many such countries generation of electricity is a local authority function. While agitating grates are popular for MSW incineration in most European countries, in the US rotary kiln systems are also used, and in Sweden and Japan fluidized bed systems contribute small proportions of total capacity.[17,18]

Data on hazardous wastes disposal and treatment are very difficult to collate, because of differing definitions in different countries, commercial confidentiality, and the varying proportions of incinerable industrial wastes in different countries. It is estimated that only 5–8% of hazardous wastes are incinerated in the OECD/Europe with some countries (*e.g.* Ireland, Spain, Greece) having no merchant capacity, others having under-capacity.[19] In-house handling of hazardous wastes which are not accounted for in national arisings figures also complicate any attempt to make comparisons between different countries. In the US, more than 90% of incinerated hazardous wastes are handled at the same facility that generated them.[20] In France, some 16% of hazardous wastes are incinerated by the merchant sector and some 14% in-house.[21] A component of hazardous wastes arisings is accounted for by contaminated soils, for which centralized soil treatment facilities, including rotary kiln incinerators, are provided in the Netherlands and Denmark and mobile plant are in use for remediation of contaminated sites in the US.[7]

3 Policy Development

Waste Management Policy

A hierarchy of waste options providing a framework for waste management forms the basis of both European Community (EC) and national policy, *i.e.*:

  (i) waste reduction at source—first priority;
 (ii) waste recycling and reuse;
(iii) recovery of raw materials and/or the energy content of the wastes;
 (iv) treatment—physical, chemical, biological, thermal—to convert wastes to a form that permits ultimate disposal; and
  (v) disposal of the residues from treatment and of other unavoidable wastes—last option. Even at this point of final disposal the objective should be to continue to utilize the inherent characteristics of the waste to

[17] International Solid Wastes Association Working Group on Waste Incineration (ISWA), 'Energy from Waste: State-of-the-Art Report', ISWA, Malmo, 1991.

[18] N. Patel and D. Edgcumbe, 'Some Observations on Municipal Solid Waste in Japan', Energy Technology Support Unit (ETSU) B00337, HMSO, London, 1992.

[19] H. Yakowitz, Proceedings of The Chemical Industry Conference, Zurich, Switzerland, March, 1991.

[20] C. R. Dempsey and E. T. Oppelt, *J. Air Waste Manage. Assoc.*, 1993, **43(1)**, 25.

[21] Anon., *HAZNEWS*, 1993, **59**, 8.

optimize reduction of its pollution potential and to extract the latent by-products (*i.e.* utilization of landfill gas).

It should be noted that in the UK a national policy on waste management has only gained any degree of transparency and structure since the publication of the Government's White Paper on the Environment in 1990.[22] The UK's traditionally decentralized and free-market approach to environmental policy and strategy development with a heavy reliance upon a private-sector based waste disposal industry (virtually 100% for hazardous wastes) has meant that both the ability and willingness to adopt, and invest in, options higher in the waste management hierarchy have been reliant almost entirely upon perceived economic benefits (such as lower liabilities and market advantage). This contrasts with the type of regime seen in the Netherlands, Denmark, and certain of the German Länder where waste disposal is controlled centrally and projections of waste arisings, required disposal and treatment capacity, and provision of facilities in terms of number and regional allocation has been planned and encouraged by central authorities. In Denmark hazardous waste is directed to a single, multi-functional, treatment facility (Kommunekemi). It should be noted that reliance upon a single facility places considerable pressure upon operation to high standards, as non-availability for any reason would significantly interrupt achievement of policy objectives.

A move to more explicit policy encouragement of the options higher in the waste hierarchy is now apparent in the UK. Waste is regarded as a renewable resource and incineration with energy recovery alongside materials recovery are stated preferred options.[23] Strategic, regulatory, action is providing some opportunity for a more structured and formal framework for consideration of the BPEO for wastes. Long-term planning is provided for in the form of: (i) waste recycling plans which have to be compiled by the Waste Collection Authorities (Section 49 of the Environmental Protection Act, 1990); (ii) waste local plans (or combined waste and minerals plans) now required to be produced by the local planning authorities under the Planning and Compensation Act, 1991; and (iii) the waste disposal plans which have to be produced by the waste regulation authorities (Section 50 of the Environmental Protection Act, 1990, replacing requirements in the Control of Pollution Act, 1974). If adequately coordinated, these plans should provide for a framework within which regional BPEOs can be formulated for various waste streams, together with a strategic reasoning that underpins the final choice of options and identification of appropriate sites. The UK Government has announced proposals for the formation of a national environmental protection agency (although this is unlikely to be formed before 1996). Such a move away from the local authority domination of waste regulation would provide the opportunity for a national strategic waste disposal plan to provide for optimum implementation of the waste management hierarchy at the national level.

[22] Department of the Environment, 'This Common Inheritance—Britain's Environmental Strategy', Cmnd. 1200, HMSO, London, 1990.
[23] Department of the Environment, 'A Review of Options', Waste Management Paper No. 1 Second Edition, HMSO, London, 1992.

The 'proximity principle', adopted both within the EC and US, requires that wastes should be handled at the nearest suitable facility to the point of arisings and complements legislative action at the EC level to minimize the transfrontier shipment of wastes to lower cost facilities. The principle raises a number of questions about the level at which appropriate facilities should be provided for particular waste streams, for example, regionally for MSW or nationally for hazardous wastes. Linked to the requirement for long-term waste disposal planning, effective implementation of the proximity principle requires good data on waste arisings, integrated planning across a number of authorities, application of BPEO principles to the identification of required options, and the willingness of local authorities to identify potential sites for facilities in the face of often strong local political pressure against.

The Environmental Protection Act, 1990, through: (i) a new emphasis on integrated pollution prevention and control; (ii) increased penalties for infringements; (iii) fees and charges for authorization of processes; (iv) introduction of a legal 'duty of care' in relation to waste management; and (v) public registers of information relating to authorizations and licences, underpins the waste management hierarchy. The resultant recognition of potential liabilities, combined with rising merchant sector disposal prices at the beginning of the 1990s, and external pressures brought about by the greater public accountability of industry in terms of environmental performance, is already having an effect upon waste producers encouraging a re-examination of processes and consideration of in-house handling of wastes currently going to landfill. This will have an effect on the amount and type of waste available for off-site disposal, for example, increased recycling is likely to generate a greater proportion of residues in the form of sludges, which could be difficult to treat. An increase in the amount of waste classed as hazardous under Directive 91/689/EEC, together with EC strategy to control such wastes, is likely to lead to greater use of incineration. Larger companies are likely to consider expansion of, or investment in, in-house treatment including incineration, although capital outlays may be significant.

Recycling Policy and Incineration

The UK's recycling policy sets a national target to recycle 50% of the recyclable component of the household waste stream by the year 2000.[23] The EC's Fifth Environment Action Programme on wastes[24] sets a target of MSW arisings of 300 kg *per capita* annum^{-1}, which is the 1985 average level, currently exceeded in most Member States, some by significant amounts, *e.g.* France at 500 kg *per capita* annum^{-1}.[25] Linked to the policies on recycling, the Action Programme does have policies on 'priority' waste streams, which for both municipal and hazardous wastes includes a ban on landfilling (*e.g.* of scrap tyres and health care wastes). At the time of writing, the UK has not translated any of the recycling

[24] Commission of the European Communities, 'Towards Sustainability—A European Community Programme of Policy and Action in Relation to the Environment and Sustainable Development', COM(92) 23 final—Vol. II, CEC, Brussels, 1992, p. 56.

[25] Organization for Economic Cooperation and Development, 'Environmental Data Compendium', OECD, Paris, 1991, p. 133.

targets specifically into targets for energy-from-waste, although these may be forthcoming if linked to the use of economic instruments to promote recycling. Even in the absence of defined targets for other waste streams it can be expected that the combined pressures identified will lead to a change in waste stream composition in the long-term.

This raises questions as to the impact on the combustible component of the stream available for incineration and also its calorific value (see Section 4). Indeed, there might appear in the UK to be the potential for conflict between recycling plans and the waste disposal plans which favour energy-from-waste provision. However, experience in other countries more advanced in recycling policy implementation than the UK (such as Japan, Sweden, the Netherlands, and the USA) indicates that high recycling rates can co-exist with high waste-to-energy rates where they are part of an integrated waste management policy.[2] A survey of experience in the USA indicates that those communities served by energy-from-waste plant have general recycling rates greater than the national average.[26] In Japan, where there is a national policy to reduce volumes requiring ultimate disposal together with a target of 80% incineration, source segregation to optimize the amount of combustible waste going for feedstock together with minimization of the non-combustibles and difficult plastic streams is seen as being compatible with materials recycling.[18] Generally, evidence suggests that the calorific value of the waste stream increases with recycling policies, but that volumes can decrease significantly. The latter could have adverse implications for the economics of energy-from-waste plant in certain locations.

4 The Economics of Incineration

Landfill versus *Incineration*

In the UK, landfill accounts for 85% of waste disposal (70% of hazardous waste).[13] An active extraction industry, availability of void space in locations relevant to waste arisings, favourable geology, the ease with which sites have been able to be 'engineered' for disposal to meet regulatory requirements, and the general availability and low price of transport, have been the key factors influencing landfill disposal prices, or 'gate fees'. Conflicts of interest in the combination of waste regulation and waste disposal functions within local authorities had the effect of keeping prices below those of the private sector, with lower regulatory and technical standards at many public sector sites. In the mid-1980s gate fees as low as £1–2 per tonne (1992 price levels) and a maximum of £8–10 would have been common. In a study for the European Commission in 1989, Environmental Resources Limited identified a ratio of landfill prices in Germany to the UK varying between 3:1 and 30:1, largely reflecting differences in technical standards between the two countries.[27]

---

[26] L. Kiser, *Warmer Bull.*, May, 1993, 8.

[27] Environmental Resources Limited, 'Charges for the Treatment and Disposal of Hazardous Waste', Report for the CEC Directorate General for the Environment, Nuclear Safety and Civil Protection, ERL, London, 1989, p. 67.

Since the late-1980s, the costs of landfill disposal have been rising. The separation of local authority regulation and disposal functions under the Environmental Protection Act, 1990, has had the impact of removing the previous conflicts of interests, and a doubling of landfill investment costs to meet rising containment engineering, gas and leachate collection, and treatment standards is being forecast.[28] These requirements are being enforced at both the UK and EC level, within the former the particular influence of the National Rivers Authority's Aquifer Protection Policy published in 1992,[29] and in relation to the latter the potential influence of the requirements proposed in the draft landfill Directive.[30] A 'new' cost in the UK will arise from implementation of Section 39 of the Environmental Protection Act, 1990, which will have the effect of extending the time over which operators of landfills have to hold a licence. With respect to the potential pollution risk to water from leachate leakage from a site, the judgement on future surrender of a licence may not be possible for many decades after a site has finished operating. Indeed, landfills could require leachate management over several centuries.[31,32] This increasing awareness of the long-term liabilities that could be held by operators will inevitably impact upon prices and also the view of landfill as the BPEO.

Landfill prices vary significantly with location, availability of suitable sites in different regions, and type of waste; however, figures relating to early 1993 suggest that for MSW, costs for disposal range from £5 to £30 per tonne, compared with MSW incineration prices between £15 and £30 per tonne, the higher figures associated with more modern facilities.[33] For industrial, clinical, and hazardous wastes incineration prices can range between £50 and £2000 per tonne dependent upon waste type. For those wastes not restricted to incineration, the availability of relatively cheap landfill space has had a major influence upon choice of option. Incineration prices for chemical wastes reflect the state of the demand and supply curve. Thus, for PCBs, where there are relatively few facilities across Europe that are able or willing to handle the wastes, prices are inflated, while for streams with a high calorific value, such as organic wastes with a low concentration of sulfur or chlorine, prices are at the bottom of the range. UK practice of co-disposal landfill (*i.e.* the joint landfill of household with commercial and industrial wastes) has had the effect of depressing prices for certain chemical wastes and providing a cheap alternative to treatment or incineration.

Several different studies have estimated that by the year 2000 UK landfill gate

[28] J.R. Holmes, 'The UK Waste Management Industry', Institute of Wastes Management, Northampton, UK, 1993.

[29] National Rivers Authority, 'Policy and Practice for the Protection of Groundwater', NRA, Bristol, UK, 1992.

[30] Commission of the European Communities, *Off. J. Eur. Communities*, 1991, **C190**, 1, and revision COM(93) 275 in *Eur. Environ.*, 1993, **413**, 1.

[31] J.D. Mather, Proceedings of the Midland Geotechnical Society Conference, The Planning and Engineering of Landfills, Birmingham, UK, July, 1991.

[32] K. Knox, Proceedings of the Energy and Environment Conference, ETSU/Department of the Environment, Bournemouth, October, 1990.

[33] Department of the Environment, 'Landfill Pricing: Correcting Possible Market Distortions', A Study by Coopers and Lybrand Ltd., HMSO, London, 1993.

fees could increase to between £10–45 per tonne (1992 prices).[12,28,34] Such prices would not by themselves be sufficient to promote a general move from landfill to incineration. Indeed an important component of the price equation is the fact that ash generation (bottom-ash from the kiln; boiler or economizer ash; and flyash comprising particulate matter in the gas stream), which could be 15–40% on a weight basis of waste incinerated, has to be landfilled to the same high standards.

The UK Government has been considering the use of a landfill levy, *i.e.* an additional charge upon the site gate fee, as an appropriate economic instrument which in addition to regulatory controls will achieve waste management objectives. A levy could be justified on a number of grounds:

   (i) To provide an incentive to recycle and recover wastes.
  (ii) To bring landfill prices into line with those elsewhere in the EC.
 (iii) To pay for the environmental costs of landfill.

The impact of a levy has proved difficult to forecast. A recent study predicts that in the long-term the most pronounced effect of a levy would be to increase the amount of waste which is incinerated, perhaps by more than 200% with a £20 per tonne levy.[33] Levies or waste disposal charges are used in other countries. For example, in the USA a number of States use them primarily for revenue-raising purposes, but also with a view to encouraging recycling. In at least thirteen States a charge is applied to both landfills and waste-to-energy plants for this reason.[34] The adoption of a levy in the UK will now depend upon policy objectives, and also decisions as to the practicalities of collection and administration to achieve these.

Incineration Capital and Investment Costs

The capital investment costs of incineration are high. Required capital investment in plant to burn 200 000 tonnes of MSW per annum to meet full EC emissions standards is estimated at about £40–45 million (1992 prices).[12,28] For industrial and hazardous waste incinerators the costs rise dramatically in the light of greater technical requirements and higher performance standards. The cost of up-grading plant to meet new emissions standards has proved too expensive, particularly where in essence the cost is in retrofitting emission control equipment to plant that may already be 15–20 years old. With over 60% of the remaining MSW incinerators operating in the UK in 1992 having a plant installed capacity of less than 100 000 tonnes, and the economically (and technically) sound scale of operation for energy purposes now regarded as being 200 000 tonnes year^{-1} minimum,[12] costs for investment are not just in up-grading but also in required replacement and expansion.

In the US, local authorities work together to 'pool' their waste arisings so as to support plant of sufficient size to keep unit costs manageable. It is noticeable that of the current private sector proposals for new plant in the UK, many are above

---

[34] Department of Trade and Industry, 'Economic Instruments and Recovery of Resources from Waste', A Study by Environmental Resources Ltd., HMSO, London, 1992.

400 000 tonnes annum^{-1} design capacity, and it has been estimated that beyond the minimum required base of 200 000 tonnes a doubling of capacity can produce a 26% decline in unit costs.[12] Clinical waste incineration has been following a similar pattern of up-grading, capacity increase, and also joint venture activities and multi-hospital facility provision following the National Health Service's loss of Crown Immunity in 1991 (*i.e.* it is now subject to direct regulation) and the designation of clinical waste incineration as a prescribed process under Part 1 of the Environmental Protection Act, 1990.

In some countries legislation has specifically recognized economic realities and either excluded small capacity plant from regulation or lowered the stringency of the emission standards. An example of the former is the USA, where 100% tyre burning facilities have been excluded from the recently enacted (1990) federal air pollution regulations. An example of the latter is EC Directive 98/429/EEC on municipal incineration in which plant are segregated into three sizes: less than 1 tonne hour^{-1}, 1–3 tonnes hour^{-1}, and greater than 3 tonnes hour^{-1}. Emission standards for key chemicals are either progressively tightened as the capacity of the plant increases (*e.g.* for dust emissions from 200 to 30 mg m^{-3}), or chemicals such as carbon monoxide are only regulated on the largest size of plant. At the time of writing there is speculation that acceptance of the proposed hazardous waste incineration Directive[35] could lead to a re-examination of these concessions for small plant, although the UK has been pressing for relaxation of the proposals to include all clinical incinerators on the basis that very small plant are often attached to residential homes and do not handle hazardous clinical waste.

To support large capital investment there is a need for a long-term (10–20 years) guaranteed waste supply and a market for any electricity. The investment costs for new plant have been largely outside of the financial capability of UK local authorities since the 1970s with no direct governmental assistance as is available in other countries. For example, in Japan central government funds the municipalities with subsidies between 25–50% of capital costs involved in plant construction, and guarantees loans on the rest of the capital.[18] Similarly in Denmark and the Netherlands there are capital allowances to support stated national objectives to recover energy from wastes. In Germany regional schemes run by the Länder authorities provide grants for investment in waste treatment plant.

Energy Recovery

Waste heat recovered from combustion gases can be used to produce steam, and if available in sufficient quantities can be used in the plant itself, in other industrial processes, for district heating schemes including residential, commercial, and leisure facilities (*e.g.* swimming pools), for electricity generation, or any combination of these. The calorific value of a substance relates to its energy potential. Coal with a value about 26–29 Gigajoules tonne^{-1} (GJ te^{-1}) compares with industrial wastes at about 16 GJ te^{-1} and MSW just under 10 GJ te^{-1}.[12] Scrap tyres and dry sewage sludge are waste streams with a higher calorific value

---

[35] Commission of the European Communities, *Off. J. Eur. Communities*, 1992, **C130**, 1.

than other wastes ($32\,GJ\,te^{-1}$ in the case of scrap tyres meaning that they can be burnt as a substitute for coal in modified conventional boilers). However, no combustion process is 100% efficient; boilers or incinerators, for example, traditionally suffer corrosion problems and so have to operate at lower steam temperatures. Expressed in terms of million tonnes of coal equivalent (tce) the total energy content of all (if it were available for burning) of the UK's wastes is estimated to be about 26–30 tce (about 15 tce of this figure from industrial wastes),[23] which represents about 10% of the UK's primary energy requirements. In the EC, industrial wastes alone are estimated to be able to provide 2–3% of the total 1987 energy consumption.[36]

As indicated in Table 2, a number of countries place great emphasis on energy recovery. In the Netherlands it is estimated that by the end of the decade MSW incineration will supply 5% of the country's energy requirements, and some cities in Germany can provide 5–10% of their current demands.[17] In Japan it has been noted that local communities who would normally oppose a new plant, have used their opposition to strengthen their bargaining position with regard to obtaining free hot water, free use of swimming pools, *etc.*[18]

In the short-term, the extent to which the UK sees a large-scale move to energy-from-waste incineration for MSW will be largely dictated by the availability of energy subsidies. Such a subsidy is provided by the Non-Fossil Fuel Obligation (NFFO) introduced under the Electricity Act, 1989, (England and Wales) which seeks to encourage use of renewable energy technologies. Schemes accepted under the NFFO receive a premium price for electricity generated. The NFFO is available until 1998. The RCEP has specifically endorsed the availability of the subsidy and recommended its extension to the whole of the UK.

A number of comparative costings have been made relative to the economics of incineration with and without energy recovery utilizing the NFFO subsidy.[12,28] These suggest that the former can result in costs per tonne of waste approximately 40% cheaper than the latter. There are a number of cost sensitivities which have to be taken into account, most particularly those of transport. Basically the lower costs of landfill means that it can tolerate greater transport costs and be competitive over longer distances. Incineration with an energy subsidy becomes competitive in relation to landfill where transfer is over approximately 100 km to a remote landfill site from the waste source (an increasingly likely scenario as suitable sites decrease and engineering costs encourage larger and fewer landfills). Despite the current limited life of the NFFO it has already proved attractive to potential investors, particularly in the private sector.[37] However, it is clear that the promotion of energy recovery as a means of achieving sustainable resource management has to take into account all of the environmental benefits and also the environmental impacts of incineration at both the strategic and site-specific levels, not least in the light of public opposition.

---

[36] K. Maniatis, 'Assessment of Incineration of Industrial Wastes: Demonstration Projects', EUR 14136 EN, Commission of the European Communities, Luxembourg, 1992.

[37] Anon., *ENDS Report*, 1992, **221**, 12.

5 Environmental Impact and Risk Assessment

Environmental Assessment

The definition of environmental impact is most appropriately adopted from the broad consideration in EC Directive 85/337/EEC. The latter requires the environmental impacts of incineration to be addressed at the stage of development consent, and a public environmental statement (ES), the outcome of an environmental assessment (EA) is statutorily required for plant which will handle hazardous waste (as defined) and may be required for other plant depending upon whether significant effects on the environment are likely. In practice, experience in the UK indicates that EAs are being undertaken for most incineration proposals in the light of the technical issues that are raised and the public concern and questioning of proposals. Over the period July 1988–December 1992 approximately 50 ESs for incinerator proposals were published.[38]

Within the UK, the requirements of Integrated Pollution Control (IPC) authorization of incinerators under Part 1 of the Environmental Protection Act, 1990, also demand that the operator provides documentation, which is made available for public comment, to show that he understands the possible effects which the plant may have on the environment, that use of the Best Available Techniques Not Entailing Excessive Cost (BATNEEC) will minimize these effects, and that where releases do occur no harm to the environment is being caused and that the BPEO is being used. There is an overlap between EA for planning permission purposes and that for authorization in the UK, although the definition of the 'environment' under IPC is limited (*e.g.* excluding the social and built environment, noise, and amenity). There is a need for rationalization of the two systems to ensure that unnecessary duplication of effort is minimized when a developer has to apply for both. However, the requirements have served to focus attention on the assessment of environmental impact from incinerators and upon the development of appropriate prediction and assessment techniques, in particular that of risk assessment (see below).

Primary attention in the literature and during public debates upon incineration has been upon the environmental impact of the deliberate and controlled releases from the process, in particular those to air. Superimposed upon these far-field (several kilometres) effects will be impacts nearer to the site boundary. Some of these impacts will result from emissions of noise, odour, dust, *etc.*, from process plant, equipment, and traffic, while others (such as the risk of fire or the threat to groundwater from spillages of waste from handling) will depend upon the intrinsic hazards of the wastes accepted at the facility. Nuisance from paper, char, and grit deposits from MSW incinerators, which in the past has been a problem as a result of low quality grates and gas-cleaning equipment, has been reduced with modern systems. Table 3 summarizes the sources and nature of potential environmental impacts from the operational phase of incineration activities that may need to be considered in EAs, and Table 4 provides an indicative comparison of the significance of impacts for different waste management options.[38]

---

[38] J. Petts and G. Eduljee, 'Environmental Impact Assessment for Waste Treatment and Disposal Facilities', John Wiley and Sons, Chichester, UK, 1994.

Sources and nature
ntial environmental
impacts from the
operational
nase of incineration

Source of impact	Nature of impact
Waste deliveries	Traffic noise; traffic impact; road accidents; dust; air pollution; unloading accidents/spillages
Plant and buildings	Visual; noise; loss of habitat through land-take; loss of visual amenity
Waste handling	Accidents/spillages; water pollution; odour; dust
Incineration	Air pollution—stack and fugitive emissions; odour; water pollution; impact on flora and fauna of emissions; human health impact (direct and indirect); visual impact of plume; fear and adverse impact upon amenity
Residue disposal	Traffic impact; fugitive dust emissions; leaching of metals and organics within the landfill

Experience in the UK indicates that the social and economic impacts of incineration and also the indirect impacts, for example, upon flora and fauna, are less openly addressed in EAs and that some proposers have had a tendency to equate the ability to operate plant to achieve required national emission standards as proof of acceptable environmental impact, with little understanding of the site-specific effects. There have also been problems where proposals represent one of several in an area, with individual EAs not able to adequately address the combined effects.[38] In the UK, adoption of the critical loads approach to EA for the facility authorization stage[39,40] emphasizes the need to predict the contribution of a single facility to the tolerable level for a pollutant in receiving media in a specific location. It could result in more stringent emission controls (including actual refusal of planning permission) than are set nationally if a local pollutant load is already unacceptable. In Japan, it has been noted that the opportunity to set local controls has been 'used' to allow siting in the face of opposition.[18] However, care needs to be taken to ensure that new developments which can operate to enhanced and effective pollution control requirements are not penalized at the expense of older facilities which cannot be closed down.

Relative Importance of Incineration as a Source of Pollutants. While public perception appears to regard incineration as a major pollutant source, work in relation to UK MSW incinerators operating in 1989–90 (*i.e.* prior to the imposition of tighter emissions standards) suggested that they might contribute less than 0.015% of total emissions to air of volatile organic compounds; less than 0.2% of SO_2, and NO_x; about 1% of CO_2; less than 2% of Cr, Cu, Ni, and Pb; and less than 3% of HCl. Cadmium and mercury contributions were estimated at approximately 24% and 13% respectively.[41] Up-dated work relating to all incinerators in 1991 produced similar figures with some revision to

[39] K. R. Bull, *Environ. Pollut.*, 1991, **69**, 105.

[40] Her Majesty's Inspectorate of Pollution, 'The Environmental Assessment of Prescribed Releases', Internal Draft for Inspectors, HMIP, London, 1992.

[41] P. Clayton *et al.*, 'Review of Municipal Solid Waste Incineration in the UK', Report LR 776(PA), Warren Spring Laboratory, Stevenage, UK, 1991.

Table 4 Matrix showing potential significance of impacts for different waste management options[1,2] (Reprinted with permission from John Wiley and Sons, Chichester, UK)

Waste management option	Odour	Health risks inhalation	Health risks foodchain	Landfill gas	Leachate	Traffic	Noise	Visual effect	Dust litter	Accidents
Incineration	XX	XXXX	XXX	—	—	X(X)	XXX	XXXX	X	XXX
Landfill	XXX	XXX	XXX	XXXX	XXXX	XXXX	XXXX	XXXX	XXXX	XXX
Landfarming	XXXX	X	XXX	—	XX	XXXX	X	XX	X	X
Composting	XX	X	XX	—	XX	XX	XX	XX	X	X
Biological treatment	XXX	X	X	—	—	X(X)	X	XX	X	X
Physico-chemical treatment	X	XX	X	—	—	X	XX	XX	X	XXX

Key: — negligible significance; X–XXXX increasing significance.
[1] J. Petts and G. Eduljee, 'Environmental Impact Assessment for Waste Treatment and Disposal Facilities', John Wiley and Sons, Chichester, UK, 1994, p. 85.
[2] The table is not intended to imply a significant impact in an absolute sense, as this will be dependent upon the location, design, and management of a particular activity.

cadmium and mercury to 32% and 11% respectively.[12] Thus, for many pollutants existing plant are seen to produce only a very small or negligible proportion of total UK emissions.

Carbon dioxide if adequately dispersed is not conventionally regarded as a pollutant relevant to health or the local environment. However, it is a contributor to the 'greenhouse' effect. While it is difficult to compare incinerator emissions with 'emissions' from landfills, as the latter cannot be so readily described in such a simple and accurate manner, a comparison between landfilling and incinerating 1 million tonnes of MSW both with energy recovery suggests that the former would generate 0.5 million tonnes of carbon as carbon dioxide while the latter would generate 0.15 million tonnes (NB the landfill calculation includes figures for carbon production as methane).[12]

PCDDs and PCDFs are some of the most well-researched pollutants in relation to incinerator emissions, in terms of total pollutant load, health risks in areas around specific sites, and occupational risks. Despite the detection of higher than background PCDD and PCDF soil concentrations in the vicinity of some incinerators[42] there has been no proof of a UK incinerator being the major contributor to local contamination. Consumption of foodstuffs such as meat, milk, fish, and eggs accounts for the primary intake of PCDDs and PCDFs, with direct inhalation constituting the rest.[43] In the Netherlands high levels in milk have been correlated with high emission rates from both MSW incinerators and other industrial sources[44] and in the UK an incinerator on an industrial facility was shut-down temporarily following identification of elevated levels in milk during routine monitoring by the Ministry of Agriculture. A specific link between the suspected source and the contamination has not been proven.[45] Several epidemiological studies have been carried out in areas surrounding incinerators in the UK following concerns about elevated levels of cancers and birth abnormalities in areas,[46,47] and the Small Area Health Statistics Unit is completing studies in areas surrounding MSW incinerators. No correlation between source and reported health defects has been proven in any case to date. US data, based on some of the most comprehensive and long-term testing, similarly demonstrate that these emissions do not present a health risk, and at hazardous wastes incinerators the most hazardous isomer (2,3,7,8-TCDD) has rarely been detected in emissions.[20]

For the first time in the UK there are now concentration limits on the release of dioxins from incinerators set by HMIP (1 ng m^{-3} limit and a 0.1 ng m^{-3} target or guide value). In 1989, it was estimated that dioxins from MSW plant may contribute about 20% of all man-made sources of dioxin in the environment (man-made sources possibly only representing 12% of the total environmental

[42] Welsh Office, 'Panteg Monitoring Report', Second Report to the Welsh Office by the Environmental Risk Assessment Unit, University of East Anglia, Welsh Office, Cardiff, 1993.
[43] ECETOC, 'Exposure of Man to Dioxins: A Perspective on Industrial Waste Incineration', Technical Report No. 49, European Centre for Ecotoxicology and Toxicology of Chemicals, Brussels, 1992.
[44] A. K. D. Liem *et al.*, *Chemosphere*, 1991, **23**, 1675.
[45] S. J. Holmes, *Environ. Prot. Bull.*, 1992, **1(2)**, 12.
[46] Scottish Office, 'Bonnybridge/Denny Morbidity Review', Scottish Office, Edinburgh, 1985.
[47] P. Elliott *et al.*, *Lancet*, 1992, **339**, 854.

load[48]), with chemical incinerators and clinical waste plant contributing 0.06% and 0.4% respectively.[41]

The RCEP report concludes that the total pollution load from modern and new plant should not be a cause for concern. Cadmium and mercury are seen as exceptions to this rule, although RCEP questioned whether more stringent standards were required in relation to these heavy metals not least because of the difficulty of measurement. RCEP concluded that there is a case for seeking to reduce the heavy metal component of the incinerator feedstock and for reducing the likelihood of leaching from solid residues (see Section 6). There is no logical reason for the setting of stricter emission standards for waste used as a fuel in energy production than for coal and oil, although this is what has happened within the EC largely for political and economic reasons. Within the waste incineration sector itself inequities in emission standards, such as in relation to boilers and industrial furnaces using wastes as supplemental fuels which are currently less stringently controlled than other incineration facilities in the UK (although not in the US), must also be carefully monitored to ensure that they are justified. A very careful and thorough environmental assessment is required to ensure effective control of all processes, while at the same time remembering that a view that incineration is a 'soft' target for pollution control even though it is not the primary source of a problem, nor a significant risk in an area, has major financial implications for waste management, and eventually society.

Risk Assessment

Risk assessment (RA) is a process in which the probability or frequency of harm for a given hazard (an event which has the potential to be harmful) is estimated. The RA proceeds through four primary stages:

   (i) the identification of the sources and components of hazards on a facility;
  (ii) determination of the release probabilities and quantities, emission or release rates and the routes or pathways by which substances could reach receptors, the fate of the substances in the environmental media through which they are transported, and the characteristics of the receptors at risk;
 (iii) estimation of risk in terms of the dose–effect relationship; and
  (iv) evaluation as to the acceptability, or tolerability, of the estimated risk.

RA is a tool to aid environmental management decisions, the output from the assessment being one of the inputs along with other material considerations (financial, social, political, technological) in any decision. RA has become an important component of the siting of new incinerators in the UK, and its use by developers within the EA process without any legislative requirement reflects recognition of its value in detailed design, understanding of the full range of impacts from both accident events and routine and fugitive emissions, and as a means of providing a more structured and open assessment for use in the public decision forum.[38] In the USA, RA has been used for specific plant, for the assessment of incineration on a national basis, and as an integral part of the

---

[48] S. Harrad and K. Jones, *Chem. Br.*, December, 1992, 1110.

development of control technology standards since the early 1980s.

Inherent uncertainties relating to data inputs, transport and fate models, and dose–response extrapolation, combined with discussion of acceptable risk criteria, provide for continuing difficulties in the presentation of RAs. However, these are not to negate the value of RA, rather they indicate the need for continuing expert attention to collation of operational and monitoring data in relation to existing plant so as to improve the quality and availability of data and information on the environmental impacts of incineration. With changing characteristics of waste streams as a result of both minimization and recovery activities and also a move from landfill to incineration, there are a number of 'untested' wastes in terms of incineration where treatability and performance testing is still required. Furthermore, as with virtually all sources of air and water pollution, the complete character of **all** compounds in emissions is not known. The expense in detecting these will have to be balanced by outstanding public questioning and the demands of the RA process.

It is noticeable that the same structured approach to assessment is only now being recognized as being valid in relation to new landfills, including consideration of liner containment failure, gas explosion hazard, and groundwater pollution risk.[49] In the past landfills have generally been designed and operated on the basis of experience and general guidance. A more rigorous assessment of landfill risks and more particularly the long-term risks over many decades may serve to provide for a more objective comparison of environmental impact between the two waste management options.

BPEO

Consideration of the BPEO involves the analysis of alternatives to determine the option that provides the most benefit or least damage to the environment as a whole, at acceptable cost. In the evaluation of options their long-term flexibility and robustness in providing for at least ten years of environmentally sound waste management is regarded as important.[23] BPEO is inextricably linked with EA (including RA), except that the former will not necessarily be determined from an EA unless the latter has specifically considered alternatives during project preparation.[38] An analysis of the BPEO should be applied in an hierarchical fashion: (i) a national study of the BPEO for different waste streams; (ii) application at the regional/local level with reference to waste arisings and the appropriate option in the light of local physical, environmental, economic, and social characteristics; and (iii) a final consideration at the site-specific level as a 'check' on the applicability of the chosen option at the particular location with environmental factors influencing the BPEO through the application of the environmental capacity approach. The establishment of the BPEO at the site level will then in turn influence strategic planning and generic policies for waste management in the longer term.

Under Integrated Pollution Control within the UK, BPEO is linked to technology and operational-based standards, *i.e.* BATNEEC (Best Available

---

[49] J. Petts, Proceedings of the Harwell Waste Management Symposium on Containment Landfill, May, 1993, Environmental Safety Centre, AEA Harwell, Oxon, UK, 1993.

Techniques Not Entailing Excessive Cost), following EC Directive 84/360/EEC on the Combating of Air Pollution from Industrial Plants. BATNEEC provides for prevention of pollution through the control of discharges to air, water, or land by appropriate process and plant design, hardware, and management systems, once the most appropriate environmental route for release or disposal of wastes (*i.e.* the BPEO) has been chosen.

Few site-specific incineration proposals have to-date explicitly linked a BPEO study with an EA in the UK, notable exceptions being in relation to sewage sludge incineration proposals, *e.g.* by Thames Water Utilities in relation to the two applications for sludge incinerators at Crossness and Beckton on the River Thames.[50] The studies, following a general BPEO methodology for sludge,[51] examined 14 principal disposal options for sewage sludge against criteria of practicability; security; environmental impacts; energy consumption; and costs. Incineration was selected as the BPEO with landfilling, digested sludge to agricultural land, and composted sludge to agricultural land being the next best options. At a strategic level a number of the English water companies have opted for incineration as the BPEO, having regard to the relative polluting potential of alternative disposal options and to the long-term security of the chosen disposal route. In contrast, three Scottish regions have decided that incineration does not represent the local BPEO, sludge re-use on land being preferred.

6 Technology Development

Alternative Technologies

In consideration of the BPEO for different waste streams, there is a requirement for continuous attention to the potential of developing and alternative technologies to provide enhanced environmental protection at acceptable costs compared to existing options. With regard to considering alternatives to the current incineration technologies key characteristics which will determine the extent of up-take relative to the total volumes of incinerable waste include:

  (i) the ability to handle large volumes of heterogeneous wastes, *i.e.* MSW;
 (ii) versatility in terms of handling both liquid and solid wastes;
(iii) energy consumption requirements and costs;
 (iv) the ability to consistently at least equal, and preferably better, any emission values associated with existing technology; and
  (v) investment requirements and commercial availability.

The potential destruction capabilities and environmental benefits of a number of alternative thermal technologies have been steadily publicized over the last decade although most are still primarily at development stage. Technologies considered to be innovative include high and low temperature plasmas, molten salt, molten glass, molten steel, and pyrolysis. The plasma arc torch is being used

---

[50] Thames Water Utilities, 'Environmental Statement—Crossness Sewage Sludge Incinerator', prepared by Ove Arup, Thames Water Utilities, Reading, UK, 1991.

[51] C. Powlesland and R. Frost, 'A Methodology for Undertaking BPEO Studies of Sewage Sludge Treatment and Disposal', Report No. PRD 2305-M/1, Water Research Centre, Medmenham, UK, 1990.

in the US and Europe for clinical wastes and for contaminated soils.[52] Other chemical and physical methods such as supercritical water, microwaves, and electrochemical processes are being explored, but are not at commercial scale.

It is often argued by opponents of incineration that decisions on new plant should be postponed in favour of more innovative technologies. However, there is considerable uncertainty as to the true cost-effectiveness of most of the new processes, since long-term, practicable, field data are not available. A careful balance is required between optimizing the BPEO for waste arisings through currently available, demonstrated, and tested options, and allocating research and demonstration resources to establishing the operability of innovative technologies.

Emission Control

The draft hazardous waste incineration directive[35] is based on progressive Best Available Technology. The guide value of $0.1\,ng\,m^{-3}$ for PCDDs and PCDFs has already been written into UK standards and Germany has adopted the standard as a mandatory limit in its Seventeenth Regulation implementing Federal law for incinerators (1990). In the Netherlands and Denmark activated carbon injection systems are in use to achieve such levels and will be used in the new plant on Teesside, north-east England. The German chemical company BASF has developed a catalytic process to abate dioxin emissions combined with removal of NO_x. A number of problems both in the standard and the gas-cleaning solutions are apparent: (i) the technical difficulties of monitoring for dioxins at such low concentrations (not least in obtaining a representative stack gas sample) mean that the limit cannot currently be measured on a consistently regular basis for compliance purposes; (ii) spent activated carbon has to be regenerated by heating which can remobilize original pollutants and its production requires considerable amounts of energy; and (iii) once standards are adopted in one Directive there may be pressure to extend them to municipal incineration which could considerably increase costs and perhaps encourage a less environmentally acceptable method of disposal.

Innovations in technology will result in new or modified processes that will gradually replace existing technologies as the BAT for a specified application provided that they are economically viable and proven at the appropriate scale. Important questions arise as to whether the emphasis should be upon achieving further reductions in emissions by installation of additional and costly gas-cleaning or rather on removing the 'problem' components of the waste feedstock (*i.e.* cadmium and mercury) and ensuring optimum operation of the combustion system to minimize dioxins.

Residue Handling

Increasing attention has been paid to the handling and disposal of the solid residues of incineration. In some countries (for example, Germany and the

[52] S. P. Howlett, S. P. Timothy, and D. Vaughan, 'Industrial Plasmas: Focusing UK Skills on Global Opportunities', Centre for Exploitation of Science and Technology, London, 1992.

Netherlands) the ash is classified as a hazardous waste, largely because of concerns over the presence of dioxins in flyash which may occlude onto ultra-fine particles and be washed out. In the USA, the ash generated has been given the same waste code as the parent waste, *i.e.* a hazardous waste produces a hazardous ash. Where so classified ash can only be disposed in suitably designed landfills equipped with leachate collection and monitoring systems of a specified standard. In Europe, flyash and bottom ash from MSW incinerators are disposed to monofill landfills (a landfill receiving only one type of waste), and segregated from biodegradable material that may mobilize heavy metals and other pollutants during degradation. In the UK, ash has been conventionally disposed with other wastes and not classified as hazardous.

Treatment of residues (in particular flyash) can transform ash into usable products, and the residues that require landfilling pose a greatly reduced hazard at the final disposal site. Several treatment processes have been examined, ranging from stabilization, solidification, and vitrification to the '3R' process, and electric smelting and heating in the absence of oxygen.[53] Vitrification produces a glass-like material suitable for landscaping or for road materials. Studies of the efficacy of solidifying or stabilizing flyash indicate that the leachate quality from the product can be one to two orders of magnitude better than the untreated ash.[54] However, solidification results in a 5–50% increase in the volume of the final product to be landfilled and there are additional process costs.

Expanded hazardous waste definitions, together with requirements to test the leachability of materials so as to classify them for disposal to landfill could provide added impetus to examination of economically attractive and feasible methods to reclaim the heavy metal components. Certainly, the UK can expect to see greater attention to solidification of flyash in the near future, which will add to the costs of incineration and also re-emphasize the importance of optimizing the combustion process through effective design and operation and analysis and management of the resultant ash, which is highly waste- and facility-specific.

7 Public Acceptance

Public Concerns About Incineration

Incinerator facilities and proposals (and also landfills) have faced opposition from local communities in most countries in Europe and North America for at least the last 10–15 years.[9] While society in general accepts the need for waste to be disposed of in a responsible and environmentally safe manner, many members of local communities amongst whom facilities are to be sited do not wish to accept potential risks and environmental disbenefits in order to relieve others of their waste problem. Exactly the same response can be witnessed in relation to many other 'locally unacceptable land-uses' (LULUs), such as motorways, airports, radioactive waste disposal, power stations, *etc.* The basis of this opposition is

[53] G. Schleger, Proceedings of the Conference Incineration the Great Debate, 18/19 February, 1992, Manchester, IBC Technical Services Ltd, London.
[54] P. H. Pardey and Graf L. Münster, in 'Energy Recovery Through Waste Combustion', ed. A. Brown, P. Evemy, and G. L. Ferrero, Elsevier Applied Science, London, 1988, p. 334.

complex and certainly not as apparently simple as the expert view that public concern is based simply upon an irrational belief of the risks to health.

There is a large literature both on the psychology of opposition to LULUs, and on the basis of the NIMBY (Not in My Back Yard) syndrome, and since the mid-1980s in particular much research effort has been directed to understanding opposition to waste facilities, both in the USA and Europe.[9,55-58] The evidence points to a complex range of factors which can be summarized in terms of the following:

(i) Perceptions of risks to health and the environment.
(ii) A lack of trust in regulatory agencies to monitor and control facilities and in the private sector to manage operations effectively.
(iii) A paucity of information availability and of communication of risk information by experts.
(iv) The exclusion of the public from fundamental policy decisions about waste management or their involvement only after initial decisions have been taken by waste managers and regulators.

Opposition to incinerators is as much a reflection of public opposition to institutional and political arrangements that appear to be directly affecting their lives, as it is to stack plumes, noise, the fear of traffic accidents handling 'toxic' waste, odour, and the general loss of amenity that any major industrial facility imposes upon a local community. As incineration facilities are rarely major local employers, adverse perceptions are compounded by a lack of any directly visible local income, except where they are 'in-house' and linked to an important local employer. In the UK, the small number of energy-from-waste plant, and the strategic nature of the chemical waste incinerators which have imported waste to support their business have served to enhance opposition. Interestingly, sewage sludge facilities and clinical incinerators within hospitals seem to have provoked less concern, although new proposals for the latter operated by the private sector have met opposition. This reflects differing perceptions of need and a greater trust in the ability of the public sector to operate plant than that in the private sector, despite the fact that it is the former MSW facilities which have been found in the past to be operating below best practice.

Improving Public Acceptance

There is a need for more open debate and greater effort to achieve a degree of informed consensus amongst interested parties. Indeed this requirement is at least equal to, if not greater than, continued technical development. A number of extended requirements for dealing with public concern can be identified relating to improving:

[55] A. Armour, *Prog. Planning*, 1991, **35**, 1.
[56] J. Petts, in 'Human Stress and the Environment', ed. J. Rose, Gordon Breach Publishers, in press.
[57] P. Wiedemann and S. Femers, *Risk Anal.*, 1991, **11(2)**, 229.
[58] K. E. Portney, 'Siting Hazardous Waste Treatment Facilities', Auburn House, New York, 1991.

(i) the decision processes for the siting of facilities and implementation of waste management strategies;

(ii) risk communication;

(iii) the management of incinerators; and

(iv) expert understanding and assessment of the risks.[9]

While most decision systems in Europe provide for public participation in decision-making for new facilities and for public information to be available relating to the control and monitoring of existing plant, far greater attention is required as to how the public can effectively be involved in the derivation of waste management policies and strategies, not least in decisions about the costs and benefits of recycling and recovery activities. There is also a need for proactive involvement much earlier in the siting process, for example, in the scoping of issues to be considered in an EA, in the derivation of site-selection criteria, and in the identification of criteria for assessing the acceptability of impacts.

The RCEP has recommended that information on the chemical composition of wastes incinerated should be made publicly available. However, in itself a long list of chemical components (even if it could be accurately defined, *e.g.* for MSW) is unlikely to assist understanding and acceptance of the processes being operated. The general availability of on-, and off-, site monitoring data, ambient environmental monitoring data, and waste input inventories (those held by both the regulatory authorities and the operator) has to be within the context of a general willingness to provide information and to liaise with local communities on an ongoing basis. An emphasis on incineration in this respect should not be at the expense of other industrial processes and a need to raise awareness of comparative pollution burdens.

Monitoring which goes beyond simple data collection for regulatory purposes should provide good long-term records and analysis sufficient to determine and predict environmental impact. Expert understanding of the negligible risks of incineration has to be proven to the public through the availability of monitoring results, full discussion of the RA methods used, and the basis of the input data (and hence the uncertainties) in any assessment. The rather restricted reference to the output of RAs and in particular the overemphasis on comparing risk figures with those for other unrelated risky activities has been one of the primary contributors to a loss of expert credibility. Finally, incineration (as with all other waste disposal and treatment activities) has to be seen to be managed effectively both in terms of operational best practice and the employment of skilled operators as well as effective regulatory control. This relates to small plant as well as large mass-burn facilities, and probably has greater financial implications for the former than the latter.

8 Conclusion

Incineration has to be discussed within the context of an integrated waste management strategy, rather than as a single option. Although landfill is unlikely to lose its prominent role in many countries, closer scrutiny of its long-term environmental impact, increasing concern amongst waste producers to protect their potential liabilities, direct restrictions on the range of acceptable wastes, and requirements for improved engineering control, are already promoting a move to incineration. The latter is technically proven as an effective waste destruction and reduction method. However, to promote incineration as an environmentally sustainable option in the public domain requires a number of actions:

(i) *Attention to the operation of integrated systems* where materials (principally metals) are segregated and recovered prior to incineration of the remainder.

(ii) *Energy recovery from all plant*, with where necessary the use of economic instruments to promote development.

(iii) *Treatment and reuse of the residues.*

(iv) *Effective and publicly accountable on-site management and regulatory control.*

(v) *Risk assessment* of all proposed and operating plant with public discussion of the results.

Most importantly, the effective management of wastes requires a long-term strategy based on a full understanding of the relative costs and benefits of different options.

Pollutants from Incineration: An Overview

P. T. WILLIAMS

1 Introduction

The disposal of waste is an increasing environmental and economic problem. Incineration of the waste is becoming a more attractive alternative than the traditional means of disposal via landfilling. Some landfill sites have problems with uncontrolled gas leakages, litter, odour, and toxic leachate. In addition, it is estimated that landfill sites close to the point of waste collection will become more difficult to acquire and will consequently necessitate higher transport costs to distant sites. When these transport costs are included together with any after care costs for the landfill site then incineration becomes an increasingly attractive waste disposal option. Incineration produces a non-putrescible and sterile ash, with a 70% reduction in mass, and 90% reduction in volume. Also, incineration has the advantage of the option of energy recovery to reduce costs. However, there is some concern that incineration of waste produces pollutants which may cause more harm to the environment than other forms of waste disposal.

The incineration of waste produces:

  (i) pollutant emissions to the atmosphere;
 (ii) contaminated waste water;
(iii) contaminated ash.

2 Pollutant Emissions to the Atmosphere

Of the pollutant emissions arising from the incineration of waste, those emitted to the atmosphere have received most attention from environmentalists and legislators. The emissions of most concern are total particulate or dust, acidic gases such as hydrogen chloride, hydrogen fluoride, and sulfur dioxide, and heavy metals such as mercury, cadmium, and lead.[1] In addition, the combustion efficiency is controlled by limits on the emission of carbon monoxide and organic carbon. Only certain countries have set limits on the emission of dioxins, whilst the European Community Directive on emissions from new municipal waste

---

[1] P. Clayton, P. Coleman, A. Leonard, A. Loader, I. Marlowe, D. Mitchell, S. Richardson, D. Scott, and M. Woodfield, 'Review of Municipal Solid Waste Incineration in the UK', Warren Spring Laboratory Report LR776(PA), HMSO, London, 1991.

Table 1 Typical emissions to the atmosphere from various incinerators $(mg\,m^{-3})^1$

Pollutant	UK (range)	Sweden Older plant (range)	Sweden Modern plant	Canada	Germany
Particulates	16–2800	1–90	1.2	—	15
CO	6–640	—	—	—	—
HCl	345–950	450–900	25	—	<2
SO_2	180–670	90–360	17	—	—
HF	—	4.5–9	<2	—	—
NO_x	—	180–360	—	—	—
Pb	0.1–50	0.45–2.7	0.06	0.055	0.358
Cd	<0.1–3.5	0.045–0.9	0.002	0.004	0.026
Hg	0.21–0.39	0.27–0.36	0.09	0.02	0.067
TCDD $ng\,m^{-3}$	0.73–1215	4.5–90	0.04	0.0	—
TCDF $ng\,m^{-3}$	6.84–1425	—	—	0.1	—
PAH $\mu g\,m^{-3}$	—	0.9–90	—	0.1	—

incinerators sets minimum combustion gas temperature, residence time, and minimum oxygen level to ensure efficient burn out.[2]

Table 1 shows a comparison of emissions to the atmosphere from certain municipal waste incinerator plants to indicate the range of emissions found.[1] These emissions data are from a wide variety of plant with different types of gas clean-up, from very simple systems to modern sophisticated systems.

Origin of Pollutants

Air pollution during waste incineration may occur in various ways:[3]

(i) Odour, dust, and litter problems may arise during the discharge, storage, and handling of waste.

(ii) The gas stream whilst passing through the waste bed may extract ash, dust, and char and carry them into the flue gas stream.

(iii) Metals and metal compounds may evaporate in the furnace to condense eventually in the colder parts of the flues and generate an aerosol of sub-micron particles.

(iv) Waste may include compounds containing chlorine, fluorine, sulfur, nitrogen, and other elements which may result in the generation of toxic or corrosive gases. Nitrogen oxides may form at the temperatures of the flame.

(v) The pyrolysis products, arising during the thermal decomposition of waste, may be combusted incompletely. These may contain CO, volatile organic compounds such as polycyclic aromatic hydrocarbons, dioxins, and furans, tar and soot particles.

[2] *Off. J. Eur. Communities*, 'Council Directive on the Prevention of Air Pollution from New Municipal Waste Incineration Plants', (89/369/EEC), Brussels, 21 June, 1989.

[3] A. Buekens and P. K. Patrick, in 'Solid waste management', ed. M. J. Suess, World Health Organization, Copenhagen, 1985, p. 79.

Odour, Dust, and Litter. The good management of the incinerator plant and discharging of the waste into the storage bunkers should ensure minimal pollution via odour, dust, and litter to the immediate surrounds of the plant. Regular sweeping and waste spillage control should be normal practice to prevent dust and litter problems. Odour may result from the waste itself and its handling, and also as an odour emitted from the stack as a product of incomplete combustion of organic waste. Incineration plants are normally kept under a slight negative pressure, because the combustion air is taken from the waste storage area, which prevents escape of odour. In addition, the waste is unloaded from transportation trucks in an enclosed building, and waste storage time should be minimized.

Odours from waste incineration are normally organic and result from incomplete combustion of organic waste material in the feed. Incomplete combustion will also result in more hazardous emissions such as dioxins and furans and a modern incinerator plant should have good, efficient combustion control to destroy any unpleasant odours. Brunner[4] has reviewed the odours from incineration of waste, threshold limit values of common odours, and their control.

Particulate Load of the Flue Gases. Particulate emissions from incinerators are the most visual to the public and can lead to complaint. The particulate is largely composed of ash; however, in addition, pollutants of a more toxic nature—such as heavy metals and dioxins and furans—are associated with particulate matter, either as individual particles or adsorbed on the surface of the particle. Acid smuts might also arise due to acidic gases such as hydrochloric, sulfuric, or even hydrofluoric acid adsorbed on the surface of soot. The dust loading of the flue gases has been shown[3] to increase with the following factors:

   (i) the ash content of the waste;
  (ii) the load factor of the incinerator;
 (iii) the amount of primary air;
 (iv) the degree of agitation of the waste;
  (v) a large degree of heterogeneity of the waste;
 (vi) too early or too late ignition;
(vii) too high or too low a grate loading;
(viii) excessive velocity of primary air;
 (ix) excessive draught;
  (x) improper balance between primary and secondary air;
 (xi) disturbance of the fire; and
(xii) excessive height of the steps between successive grate sections.

The design of the incinerator also influences the rate of particulate loading in the flue gases. Relevant factors include the incinerator size, grate type, and the combustion chamber design. Larger incinerator units seem to have slightly higher emissions rates. Part of the increase is due to the higher rates, and hence higher underfire air rates, for the larger size, possibly due to the consequence of

[4] C. R. Brunner, 'Hazardous Air Emissions from Incineration', Chapman and Hall, New York, 1985.

the higher natural convection currents encountered in larger units. The emission rates from reciprocating grates have been found to be higher than from other grate types. Particulate emissions can also be reduced substantially by use of multichambers and the use of low-arch combustion chambers.

When carbon-containing wastes are combusted in conditions of high temperature and low oxygen content this can lead to the formation of soot. Polycyclic aromatic hydrocarbons (PAH) have been cited as intermediaries in soot formation, an alternative proposed pathway is via acetylene. The control of soot formation is via adequate residence time for the combustion process to completely burn out any soot being formed, with good mixing of the primary and secondary combustion air.

The emission of untreated flue gases would give rise to a dark plume and the deposition of dust downwind of the incinerator stack. The size range of incinerator particulates is from $<1\,\mu m$ to $75\,\mu m$,[5] larger particles settling out prior to the flue. It is the ultrafine particles that are of particular concern since these are composed of ash, fine heavy metal particles, and organic particles which, because of their size, can pass deep into the respiratory system of humans.

The emissions are controlled by dust collection systems such as mechanical separators, wet scrubbers, or fabric filters. Larger particles, $15-75\,\mu m$ are effectively removed by cyclones with up to 85% efficiency; the removal of finer particles is achieved with either fabric filters or electrostatic precipitators. In some cases these may be required to be preceded with a dry scrubbing system to capture the ultrafine particles. In most modern gas clean-up systems combinations of different systems are required to meet the legislative requirements placed on incinerator emissions.

Evaporation of Metals. Metals and metal compounds are present in the components of raw waste. For example, municipal refuse may contain lead from lead-based paints, mercury and cadmium from batteries, aluminium foil, lead plumbing, zinc sheets, volatile salts, *etc*.

Table 2 shows the range of trace components found in municipal solid waste from various countries in Europe.[6,7] High levels occur and the concentrations are very variable. The extent of evaporation of these metals and metal compounds in the furnace depends on complex and interrelated factors such as operating temperature, oxidative or reductive conditions, and the presence of scavengers, mainly halogens.[3] These metals and salts are relatively volatile and have low boiling points, for example; Cd, b.p. 765 °C; Hg, b.p. 357 °C; As, b.p. 130 °C; $PbCl_2$, b.p. 950 °C; and $HgCl_2$, b.p. 302 °C; however, for some compounds the temperatures are not known. Of the heavy metals, cadmium, mercury, and lead are deemed of most importance in relation to municipal waste incinerators, since, although other metals occur, their toxicities or emission levels are much lower. The speciation of the metals in the incinerator off-gas is strongly influenced by the presence of compounds of chlorine, sulfur, carbon, nitrogen,

[5] W. R. Niessen, 'Combustion and Incineration Processes', Marcel Dekker Inc., New York, 1978.
[6] K. E. Lorber, in 'Sorting of Household Waste and Thermal Treatment of Waste', ed. M. P. Ferranti and G. L. Ferrero, CEC Elsevier Applied Sciences, Essex, 1985.
[7] S. L. Law and G. E. Gordon, *Environ. Sci. Technol.*, 1979, **13**, 432.

The range of trace components in municipal waste (g tonne^{-1})[6,7]

Trace component	USA[a]	Europe[b]
Fe	1000–3500	25 000–75 000
Cr	20–100	100–450
Ni	9–90	50–200
Cu	80–900	450–2500
Zn	200–2500	900–3500
Pb	110–1500	750–2500
Cd	2–22	10–40
Hg	0.7–1.9	2–7

[a]Ref. 7.
[b]Ref. 6.

fluorine, and others during combustion and gas cooling. The off-gas contain metals and chlorine species, particularly hydrogen chloride which leads to the formation of metal chlorides. For example, Vogg et al.[8] have shown that cadmium is easily volatilized during incineration and is oxidized in the presence of hydrogen chloride to cadmium chloride as the main product. They also found that 30% of the cadmium remains in the slag whereas 70% occurs in the furnace off-gas; 99% of the cadmium was shown to occur as a condensate on associated dust particles and only 1% was present in the gas phase. Mercury has also been shown to be present largely in the halogenated form, predominantly mercury(II) chloride and to a lesser extent mercury(I) chloride. Whilst initially mercury is vaporized as the metal in the furnace, it quickly becomes oxidized to the halogenated form and only a small percentage is present as metal vapour.

The distribution of the metals in various outputs from the incinerator have been investigated by a number of workers; for example, Brunner and Monch,[9] Buekens and Patrick,[3] and Carlsson.[10] Brunner and Monch[9] show the distribution of heavy metals as a mass balance into and out of an incinerator equipped with an electrostatic precipitator as the only gas clean-up measure in terms of that fraction emitted to the flue gas, that captured in the electrostatic precipitator, and in the slag from the furnace.

Figure 1 shows the partitioning of iron, copper, zinc, lead, cadmium, and mercury from municipal waste incineration, in the flue gas, electrostatic precipitator ash, and in the furnace bottom slag.[9] It is suggested that the partitioning is a function of the physico-chemical properties of the elements and their derived compounds, such that volatile mercury and cadmium compounds with high vapour pressures and low boiling points are most likely to be found in the flue gas. Metals with a low vapour pressure, such as lead and zinc, are retained better in the slag and are less concentrated in the electrostatic precipitator dust. Iron is almost completely trapped in the slag, whilst the slightly more volatile copper shows a similar behaviour. Buekens and Patrick[3] have also shown that where the gas clean-up system consists of an electrostatic precipitator the majority of the metals are collected in the bottom ash, clinker fraction. A major

[8] H. Vogg, H. Braun, M. Metzger, and J. Schneider, *Waste Manage. Res.*, 1986, **4**, 65.
[9] P. H. Brunner and H. Monch, *Waste Manage. Res.*, 1986, **4**, 105.
[10] K. Carlsson, *Waste Manage. Res.*, 1986, **4**, 15.

Figure 1 The partitioning of metals by municipal solid waste incineration, to the flue gas, electrostatic precipitator dust, and bottom slag. (Figure (ref. 9) reproduced with permission from *Waste Manage. Res.*, 1986, **4**, 105.)

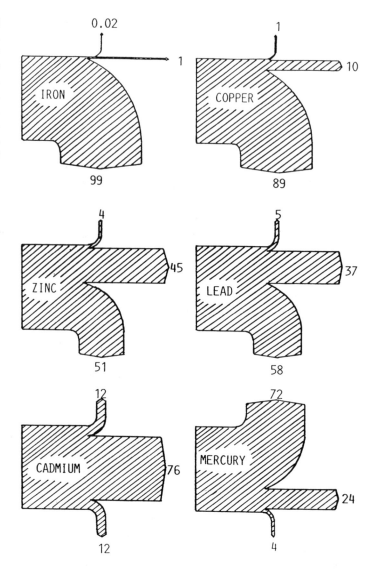

part of the airborne metal (flyash) is collected by the electrostatic precipitator and only a minor part escapes collection as the fly dust. However, the more volatile metals such as mercury, cadmium, and to some extent, lead, and zinc remain uncollected in the electrostatic precipitators as significant concentrations in the emitted flue gases. Consequently, more sophisticated emissions control equipment is required to trap the more volatile heavy metals, particularly mercury. Carlsson[10] has compared different gas cleaning systems from municipal waste incinerators in Switzerland, Germany, and two Swedish plants with particular regard to the heavy metal emissions. The four clean-up systems consisted of an electrostatic precipitator and electrostatic scrubber, an electrostatic precipitator and scrubber with condensation, a spray tower of hydrated lime slurry and an electrostatic precipitator, and dry injection of hydrated lime with a fabric filter. The highest collection efficiency for the metals under consideration was the dry injection of hydrated lime and fabric filter system. This is due to the high efficiency in collecting sub-micron particles and also for gaseous mercury compounds. Reimann[11,12] has also shown that scrubbing of the flue gas with a lime slurry after the electrostatic precipitator is effective in removing over 99% of most heavy metals. The exception is the more volatile mercury, for which only 90% removal occurred because of the high vapour pressure of the derived mercury compounds. Bergstrom[13] has suggested that lime injection with flyash prior to the fabric filters causes a marked increase in the collection efficiency for mercury; the mercury becomes bound effectively to the flyash and becomes trapped by the fabric filter. Even so, collection efficiencies of only 89% were reported. Similarly, Buekens and Schoeter[14] showed for a German municipal waste incinerator that 90% removal of mercury could be achieved at 60–70 °C using a strongly acidic wet scrubbing liquor (pH 1–3) which was subsequently removed as sludge from the scrubbing water. Since vapour condensation is so important for mercury control, cooling the flue gases to below 160 °C increases the collection efficiency for mercury. Carlsson[10] also showed for a range of different gas clean-up systems that the highest removal rates only reached 85% for mercury where dry lime injection with fabric filters was used; however the collection efficiency for cadmium, lead, and zinc was over 99.5%. However, in a later paper, Carlsson[15] showed that improvements in technology for dry scrubbing with fabric filters could improve the efficiency of collection for mercury. He reported work from Canada and Sweden using dry lime scrubbing followed by fabric filters which gave removal efficiencies for mercury of 97% and 99%. The collection efficiencies for cadmium, lead, and zinc, and other heavy metals were over 99.8%. To meet the legislative requirements for mercury emissions some form of reagent addition may be necessary upstream of the scrubber system and fabric filter. Additives which have proved effective are sodium sulfide, TMT 15 (trimercapto-*s*-triazine), and activated carbon. The additive is added at concentration levels of between 0.1 and $0.5 \, \mathrm{g \, m^{-3}}$ of waste gas and high removal efficiencies have been reported. The additives add to the cost of gas clean-up with sodium sulfide being most cost effective, followed by activated carbon which is approximately three times the cost and TMT 15 seven times the cost of sodium sulfide. Activated carbon also has the advantage of removing dioxins and furans from the gas stream. An alternative to gas clean-up of heavy metals is to eliminate them from the raw waste material. The recycling of batteries for the removal of cadmium and mercury has been shown to be effective in reducing the emissions of these metals.[16]

The heavy metals are associated with the particulate, since volatilization of

[11] D. O. Reimann, *Waste Manage. Res.*, 1986, **4**, 45.

[12] D. O. Reimann, *Waste Manage. Res.*, 1989, **7**, 57.

[13] J. G. T. Bergstrom, *Waste Manage. Res.*, 1986, **4**, 57.

[14] A. Buekens and J. Schoeters, 'Thermal Methods in Waste Disposal'. Study performed for EEC under contract number ECI 1011/B7210/83/B, Brussels, 1984.

[15] K. B. Carlsson, in 'Energy Recovery Through Waste Combustion', ed. A. Brown, P. Evemy, and G. L. Ferrero, Elsevier Applied Science, Essex, 1988.

[16] Anon., 'Heavy metal and PAH compounds from municipal incinerators', Environmental Health Series No. 32, World Health Organization, Copenhagen, 1990.

metals occurs during the combustion of the waste and subsequent condensation at lower temperatures and adsorption onto the fine particulates in the flue gas. It has been shown that there is an increasing concentration of metals with decreasing particle size for municipal waste incinerators.[17,18] A large fraction of the particles are less than 10 μm in size, and consequently of respirable size which can easily be ingested. In addition, their small size promotes both short and long-range dispersion, which has been demonstrated for municipal waste incinerators.[19]

Heavy metals exert a range of toxic effects including neurological, hepatic, renal, and hematopoietic.[19] The effects have been reviewed by Friberg *et al.*[20] Major episodes of ill health have been reported among populations acutely and chronically exposed to heavy metals, particularly cadmium, mercury, and lead. The presence of cadmium represents a health risk via accumulation in living tissue leading to respiratory ailments, kidney damage, hypertension, and, in extreme circumstances, damage to bones and joints.[21] Mercury and its compounds give rise to toxic effects associated with the central nervous system, the major areas affected being associated with the sensory, visual, and auditory functions, as well as those concerned with coordination.[16] Lead exposure has been associated with disfunction in the haematological system and the central nervous system. Decreases in intelligence and abnormal behaviour have been reported in children subject to exposure of increased levels of lead although such effects are currently controversial.[16] The primary route for human exposure to heavy metals released by incineration is the food chain. However, it has been concluded that no effects on health have been linked to the release of heavy metals from incineration plants.[22]

It is difficult to compare emissions from incinerators with fossil fuel combustion plants since the plants are markedly different, incorporate different gas clean-up systems, and obviously have different heavy metal inputs. However, emissions of heavy metals from coal-fired power stations can be presented for comparison with data in relation to incinerators.

Table 3 shows data for the concentration of metals in particulates emitted as fly-dust from the stack of a municipal waste incinerator in comparison to a coal fired power plant.[14] Clearly the levels of cadmium, lead, mercury, and zinc are much higher from waste incineration than from the coal plant. However, levels of other metals (noticeably strontium and vanadium) are higher from the coal power plant.

As is the case for municipal waste incinerators, it has also been shown that the majority of the fly-dust released from coal-fired power stations is in the respirable size range of less than 10 μm. In addition, the metals are more predominant in the

[17] B. Bouscaren, in 'Energy Recovery Through Waste Combustion', ed. A. Brown, P. Evemy, and G. L. Ferrero, Elsevier Applied Science, Essex, 1988.

[18] R. R. Greenberg, W. H. Zoller, and G. E. Gordon, *Environ. Sci. Technol.*, 1978, **12**, 566.

[19] R. A. Denison and E. K. Silbergeld, *Risk Anal.*, 1988, **8**, 343.

[20] L. Friberg, G. F. Nordberg, and V. B. Vouk, 'Handbook of the Toxicology of Metals, Vol. I', Elsevier, Amsterdam, 1986.

[21] S. D. Probert, K. Kerr, and J. Brown, *Appl. Energy*, 1987, **27**, 89.

[22] Royal Commission on Environmental Pollution, 17th Report, 'Incineration of Waste', HMSO, London, 1993.

	Municipal waste incinerator	Coal fired power station
As	180	490
Ba	2100	1900
Be	4	30
Cd	500	30
Cr	650	370
Co	140	40
Cu	1450	300
Pb	20 000	2100
Hg	>130	5
Sr	290	1800
V	160	850
Zn	48 000	2800

le 3 Comparison of
:rator and coal fired
r station trace metal
rations in emitted fly
dust (mg kg^{-1})[14]

fine fraction of the particles and are concentrated at or near the surface of the particles.[23,24]

Contaminants Containing Cl, F, S, or N. Typical waste contains about 7000–8000 mg kg^{-1} chlorine, 100–200 mg kg^{-1} fluorine, and 2700–5000 mg kg^{-1} sulfur.[11] The waste chloride and fluoride are in the form of waste plastics, for example, PVC and PTFE, in paper and board, and as sodium chloride. The sulfur content is low compared to coal sulfur contents; however, some waste oils may contain up to 5% sulfur and 1–2% chlorine.[21]

Normally because the combustion of the volatile fraction in an incinerator is almost complete the flue gases consist mainly of nitrogen, oxygen, and carbon dioxide. Combustible waste compounds which contain the elements chlorine, fluorine, sulfur, or nitrogen during combustion generate gaseous contaminants such as hydrogen chloride, hydrogen fluoride, sulfur oxides, and nitrogen oxides:

$$C, H, Cl, F, S, N + O_2 \rightarrow CO_2 + H_2O + HCl + HF + SO_2 + NO$$

At the temperature of the flame CO_2 and H_2O may partially dissociate, but the resulting CO, H_2, and O_2 recombine when the temperature decreases; many of the other products (*e.g.* HCl, HF, and SO_2) are stable.

At lower temperatures SO_2 and HCl can be thermally or catalytically oxidized, *e.g.*

$$2SO_2 + O_2 \rightarrow 2SO_3$$
$$4HCl + O_2 \rightarrow 2H_2O + 2Cl_2$$

However, the conversion of SO_2 and HCl remains very limited unless catalytic dust particles are present. Also Cl_2 gas is reduced by numerous gases or solid reducing agents, *e.g.*

$$Cl_2 + SO_2 + H_2O \rightarrow 2HCl + SO_3$$

Chlorine is not normally detectable in the furnace emissions.

[23] C. L. Fisher, B. A. Prentice, D. Siberman, J. M. Ondor, A. H. Biermann, R. C. Ragaini, and A. R. McFarland, *Environ. Sci. Technol.*, 1978, **12**, 447.
[24] R. W. Linton, A. Loh, D. F. S. Natusch, C. A. Evans, and P. Williams, *Science*, 1976, **191**, 852.

P. T. Williams

The origin of HCl in incinerator flue gas has been the subject of much research due to the corrosive nature of HCl at low temperature (*i.e.* dew point corrosion) and high temperature corrosion when it dissolves in molten salts. The major source of HCl is regarded as PVC plastic, and a direct relationship between HCl in the flue gas and PVC content of the waste for a municipal waste incinerator has been demonstrated.[14] However, other sources such as metal chlorides like NaCl or $CaCl_2$, derived from paper, board, and vegetable matter, also are regarded as sources of HCl.[25] PVC emits HCl by a gradual process of thermal decomposition which takes place between 180 and 600 °C. Buekens and Schoeters[14] have shown that 60% of HCl is due to PVC, 36% due to paper products, and 4% to grass and leaves in laboratory studies of incineration.

Reported emissions of HCl from coal-burning plants in the USA have shown levels of between 14 and 220 p.p.m. whilst incinerator emissions have shown levels of between 215 and 1250 p.p.m.[26] The data are somewhat misleading since they will depend on capacity; total mass emissions per year would be more representative, but they give some idea of the equivalent levels from large scale sources of pollution. Notably, the equivalent SO_2 emissions were between 127 and 446 p.p.m. for the coal plant and between 21 and 73 p.p.m. for the incinerator. Municipal incinerators are regarded as only a minor source of SO_2 emission when compared to power plants and industrial boilers firing heavy fuel oil or coal.[1] The higher HCl and lower SO_2 emissions from incinerators has prompted manufacturers to suggest a better measure of incinerator emissions would be total acidity, combining the acidic gases and thus representing a better comparison with other forms of fuel.

Hydrogen fluoride is even more reactive and corrosive than HCl and arises from combustion of fluorinated hydrocarbons. Typical emission levels of between 3 and 5 mg m^{-3} are reported as average values. HF can be controlled by scrubbing of the flue gas.

Nitrogen oxides (NO_x arise from the nitrogen in the fuel and by direct combination of the atmospheric nitrogen and oxygen present; this occurs more rapidly at high temperature. In practice, thermal NO is formed almost exclusively in the flame, particularly under oxidizing conditions; in reducing conditions little NO is formed.[3] NO_x generation is increased with high nitrogen-content of the waste, and high flame temperatures. NO_x generation is reduced by using either low temperature combustion, or high temperature combustion under reducing conditions. Recirculation of the flue gases and addition of ammonia also are known to reduce NO_x generation.[3] At temperatures below 200 °C, NO is slowly oxidized into NO_2. Oxidation reactions continue after emission of the flue gases into the atmosphere. Levels of NO_x reported from USA incinerator and coal-burning plants show similar emission levels of about 100–200 p.p.m.

Chlorine, fluorine, sulfur, and nitrogen may also occur in the ash of the waste as bottom ash, flyash, or dust if they are present in the form of thermally stable compounds, or incorporated by adsorption and reaction of the HCl, HF, SO_2, *etc.*, with metal oxides and hydroxides present in the ash as a further source.

[25] S. Uchida, H. Kamo, and H. Kubota, *Ind. Eng. Chem. Res.*, 1988, **27**, 2188.
[26] C. Parker and T. Roberts, 'Energy from Waste: An Evaluation of Conservation Technologies', Elsevier Applied Science, London, 1985.

The emission of the aforementioned pollutant gases to the atmosphere contributes to the well-documented acid rain with its associated environmental damage. NO, after atmospheric oxidation to NO_2, is active in the generation of photochemical smog. Condensation of the acid gases at temperatures below the acid dew point, approximately 140 °C, produce corrosive damage to the back-end incinerator plant. The dew point for HCl is lower than that for SO_2, falling between 27 °C and 60 °C depending on the HCl content and water content of the flue gas.[27] Krause[27] has also shown that HCl is important in high temperature corrosion of metal surfaces such as heat exchangers. High temperature corrosion involves a series of interactions between metal, scale deposits, slag deposits, and flue gases. The rate of corrosion is influenced by temperature, the presence of low melting-point phases such as alkali bisulfates and pyrosulfates, HCl, SO_3, the nature of the metal, and the periodic occurrence of reducing conditions. The low-melting phases are eutectic mixtures formed between metal salts and the metal surface, with metal chlorides as the most likely source of molten salt corrosion because of their low melting points.

Brna[28] has reviewed clean-up of flue gases including acid gas control from municipal waste incinerators. Dry, semi-dry, and wet processes are used to remove acid gases produced by waste combustion. Dry systems use a dry powder and possibly up-stream humidification to improve gas/sorbent reaction. Semi-dry processes use an alkaline sorbent slurry or solution which is atomized into fine droplets and injected into the flue gas; the droplets react and dry in the hot flue gases to produce a dry powder. Adsorption is improved by the use of a downstream fabric filter which increases contact time between the gases and the alkaline filter cake formed on the filter by the adsorbent. Wet scrubbing systems use slurries and solutions at lower temperatures than the semi-dry system and produce a wet solid or sludge reaction product. The adsorbents used include, CaO, $Ca(OH)_2$, and NaOH for wet scrubbers. Brna[28] has reported removal efficiencies of 90% for HCl and > 70% for SO_2, for a municipal waste combustor using CaO spray absorbers followed by either fabric filters or electrostatic precipitators. In a different combustor, where three CaO spray dryer absorber/fabric filters units were used, the results indicated 99% removal of HCl and 93–98% removal of SO_2. These systems were also effective for removal of HF.

Oxides of nitrogen cannot be reduced by scrubbing because of their low solubility; the NO_x are largely present as NO. Additives such as sodium chlorite added to the scrubber oxidize the NO to NO_2 which is then more soluble in the down-stream scrubber units. The addition of ammonia to form nitrogen and water may be also a solution and NO_x reductions of over 60% have been reported.[28]

Products of Incomplete Combustion. The volatile matter arising from incineration of waste is normally completely combusted by providing adequate residence time, post-combustion temperature, and turbulent mixing. The concentration level of carbon monoxide then consistently remains below 0.1 volume %. Incomplete combustion may occur when the incinerator is improperly operated;

[27] H. H. Krause, in 'Incinerating Municipal and Industrial Waste', ed. R. W. Bryers, Hemisphere Pubs. Corp., New York, 1991.
[28] T. G. Brna, *Combust. Sci. Technol.*, 1990, **74**, 83.

for example, operation at excessively low temperatures (below 800 °C) or overloading of the plant. The occurrence of incomplete combustion can be detected by monitoring the flue gas composition. The most contentious products of incomplete combustion from the incineration of waste are polycyclic aromatic compounds, dioxins, and furans.

(a) *Polycyclic Aromatic Compounds or PAC* are compounds based on aromatic benzene rings which are fused to form two or more polycyclic rings. Within the PAC class sub-classes exist; for example, polycyclic aromatic hydrocarbons (PAH, in older texts sometimes called PNAH or PNAs) where there are no heterocyclic atoms in the ring; sulfur-containing PAH (PASH); nitrogen-containing PAH (PANH); nitro-containing PAH (NPAH), *etc.*

Polycyclic aromatic compounds are known to occur naturally in the environment; for example in sediments, in fossil fuels, and—from natural combustion—in forest fires. The major sources of PAC, however, are anthropogenic; examples include oil- and coal-fired power generation plant, coke production, residential furnaces, in diesel and gasoline engines, and—of most relevance here—in waste combustion.[29,30] Concern over the emission of PACs to the environment is centred on the associated health hazard, because PACs comprise the largest group of carcinogens among the environmental chemical groups.[29] PACs absorbed into airborne particles are believed to be a major contributory reason why death rates from lung cancer are higher in urban than in rural areas.[31] Cancers of the lung, stomach, kidneys, scrotum, and liver have been associated with exposure to PACs.[29,32] Not all the large number of PACs known to exist are biologically active, however, and many have not been tested either individually or as they occur in complex mixtures. The relative carcinogenic activity of some PACs have been assessed, and many reviews exist on the health hazard associated with PACs.[29,32] In addition, PACs are known to form in the combustion process and have been suggested as precursors to the formation of soot in combustion systems.

Table 4 shows the PAC emission from incinerators firing municipal solid waste and refuse-derived fuel (RDF).[26,33,34] In the Eksjo incinerator in Sweden, using RDF pellets in a fluidized bed,[26] a large percentage of the PAC were reported to be adsorbed on the surface of the flyash. Table 4 also shows the distribution of PAC from the stack gases from a UK incinerator.[33] It was also reported that the largest emission was associated with the solid residues associated with the flyash and clinker, the cleaned flue gas ranking second. The water concentration was low, reflecting the low solubility of PAC in water. Also shown in Table 4 are PAC emissions during the cold start-up of a municipal solid waste incinerator.[34] The emissions are clearly much higher, reflecting the less than optimum combustion efficiency during start-up. Under normal operation the concentration of individual PACs never exceeded $10 \, \text{ng m}^{-3}$. Also detected were a number of

[29] M. L. Lee, M. Novotny, and K. D. Bartle, 'Analytical Chemistry of Polycyclic Aromatic Compounds', Academic Press, New York, 1981.

[30] P. T. Williams, *J. Inst. Energy*, 1990, **63**, 22.

[31] C. B. Love and E. P. Seskin, *Science*, 1970, **169**, 723.

[32] E. Gelboin and S. Tso, 'Polycyclic Hydrocarbons and Cancer', Academic Press, New York, 1978.

[33] I. W. Davies, R. M. Harrison, R. Perry, D. Ratnayaka, and R. A. Wellings, *Environ. Sci. Technol.*, 1976, **10**, 451.

[34] A. C. Colmsjo, Y. U. Zebuhr, and C. E. Ostman, *Atmos. Environ.*, 1986, **20**, 2279.

Table 4 Emissions of polycyclic aromatic compounds from municipal waste $(\mu g\,m^{-3})^{26,33,34}$

PAC	UK[a]	Sweden[b] RDF	Sweden[c] MSW
Fluorene	—	0.2	12.0
Methylfluorenes	—	0.09	—
Phenanthrene	—	—	43.0
Carbazole	—	0.05	—
Fluoranthene	0.58	0.2	11.0
Pyrene	1.58	0.09	6.8
Benzo(*a*)anthracene	} 0.72	0.1	1.1
Chrysene		0.5	3.0
Benzofluoranthenes	0.32	0.04	4.0
Benzo(*a*)pyrene	} 0.02	0.04	—
Benzo(*e*)pyrene		0.04	0.7
Perylene	0.18	0.04	—
Benzo(*ghi*)perylene	0.42	0.04	—
Coronene	0.04	—	—

[a]Ref. 33.
[b]Ref. 26.
[c]Ref. 34.
RDF: Refuse derived fuel.
MSW: Municipal solid waste.

halogenated PACs, chlorobenzenes, and chlorophenols which are known to act as precursors for the formation of dioxins and furans.

The PACs reported include some species known to be biologically active in human and bacterial cell tests; for example benzo(*a*)pyrene, benzo(*e*)pyrene,[29] phenanthrene, methylphenanthrenes, and fluoranthene,[35] and the methyl-fluorenes.[36]

The levels of PAC reported for incinerator emissions are very low when compared with other emission sources of PAC; for example, Table 5 shows PAC emissions from coal- and oil-fired power stations[37] and a diesel engine,[38] reflecting the combustion process in all its forms as a source of PAC. Diesel engine exhaust in particular contains high levels of PAC, orders of magnitude higher than those reported from the RDF, and municipal waste incinerators.

The physical and chemical properties of PACs suggests that control mechanisms introduced for the control of dioxins and furans would easily control PAC emissions from incinerators. For example, Davies *et al.*[33] have shown that gas clean-up systems incorporating sprayers and electrostatic precipitators have a high efficiency for removal of PAC.

(b) *Polychlorinated dibenzo-*p-*dioxins (PCDD) or 'dioxins' and the closely related*

[35] J. P. Longwell, in 'Soot in Combustion Systems and its Toxic Properties', ed. J. L. Lahaye and G. Prado, Plenum Press, New York, 1983, p. 37.
[36] T. R. Barfnecht, B. M. Andon, W. G. Thilley, and R. A. Hites, in 'Proceedings of the Fifth International Symposium on PAH', Columbus, OH, 1980.
[37] P. Mascelet, M. A. Bresson, and G. Mouvier, *Fuel*, 1987, **66**, 556.
[38] P. T. Williams, K. D. Bartle, and G. E. Andrews, *Fuel*, 1986, **65**, 1150.

Table 5 Emissions of polycyclic aromatic compounds from fossil fuel combustion (μg m^{-3})[37,38]

PAC	Diesel engine[a]	Coal fired power station[b]	Oil fired power station[b]
Fluorene	13	0.4	0.6
Phenanthrene	80	4.4	2.5
Methylphenanthrenes	215	—	—
Fluoranthene	65	2.5	0.9
Pyrene	42	6.4	0.5
Benzo(a)anthracene	26	0.4	0.1
Benzo(a)pyrene	13	—	0.05
Benzo(e)pyrene	7	0.02	0.01

[a]Ref. 38.
[b]Ref. 37.

polychlorinated dibenzofurans (*PCDF*) *or 'furans'* constitute a group of chemicals that have been demonstrated to occur ubiquitously in the environment. They have been detected in soils and sediments, rivers and lakes, chemical formulations and wastes, herbicides, hazardous waste site samples, landfill sludges, and leachates.[39] PCDD and PCDF have a number of recognized sources, among which are their formation as by-products of chemical processes such as the manufacture of wood preservatives and herbicides, the smelting of copper and scrap metal, the recovery of plastic-coated wire, and natural combustion such as forest fires.[40,41] More contentiously, they are found in combustion products, the ash, stack effluents, water, and other process fluids from the combustion of municipal waste, coal, wood, and industrial waste.[39] The concern over dioxins and furans arises from a number of animal studies which show that for some species they are highly toxic at very low levels of exposure.[42,43] The extrapolation of these animal data to man, though contentious, has led to dioxins and furans acquiring their notoriety as 'the most toxic chemical known to man'. PCDD and PCDF are highly stable environmentally and present difficult sampling and analytical problems because of interferences, their low concentration, and their perceived toxicity.[42–44] PCDDs and PCDFs have been involved in a number of incidents in recent years which give them their notoriety. For example, the Seveso accident in Italy, in 1976, the herbicide spraying program of Agent Orange in Vietnam in the late 1960s, and the Times Beach, Missouri, land poisoning of 1982.

The generalized molecular structures of PCDD and PCDF are shown in Figure 2; they are tricyclic aromatic compounds containing two (dioxin) or one

[39] T. O. Tiernan, in 'Chlorinated Dioxins and Dibenzofurans in the Total Environment', ed. G. Choudhary, L. M. Keith, and C. Rappe, Butterworth, London, 1983, p. 211.
[40] N. Steisel, R. Morris, and M. J. Clarke, *Waste Manage. Res.*, 1987, **5**, 381.
[41] Department of the Environment, Pollution Paper No. 27, 'Dioxins in the Environment', HMSO, London, 1989.
[42] H. Tosine, in 'Chlorinated Dioxins and Dibenzofurans in the Total Environment', ed. G. Choudhary, L. M. Keith, and C. Rappe, Butterworth, London, 1983.
[43] D. Oakland, *Filtr. Sep.*, Jan/Feb, 1988, 39.
[44] P. T. Williams, *J. Inst. Energy*, 1992, **65**, 46.

Dioxin and furan
molecules and the
3,7,8-tetra isomers

Dioxin molecule

Furan molecule

2,3,7,8-Tetrachlorodibenzo-*p*-dioxin

2,3,7,8-Tetrachlorodibenzofuran

(furan) oxygen atoms. Each of these structures represents a whole series of discrete compounds having between one and eight chlorine atoms attached to the ring; for example, Figure 2 shows the tetra-isomers, with four chlorine atoms in the 2, 3, 7, and 8 positions, *i.e.* 2,3,7,8-tetrachlorodibenzo-*p*-dioxin (2,3,7,8-TCDD) and 2,3,7,8-tetrachlorodibenzofuran (2,3,7,8-TCDF). Since each chlorine atom can occupy any of the eight available ring positions it can be calculated that there are 75 PCDD isomers and 135 PCDF isomers. All the PCDDs and PCDFs are solids with high melting and boiling points, and with low solubility in water. Many of these isomers have not been prepared in pure form and hence their toxicology has not been assessed and their identification is difficult.

The potential threat of PCDDs and PCDFs to humans should be assessed bearing in mind that the 75 PCDD isomers and 135 PCDF isomers have differing toxicities and are often present in multiple mixtures of the isomers. The toxicities of the PCDF isomers generally parallel those of PCDD.[39] The assessment of the toxicity of PCDD and PCDF mixtures has led to the development of the Toxic Equivalent (TEQ) scheme. This uses the available toxicological and biological data to generate a set of weighting factors each of which expresses the toxicity of a particular PCDD or PCDF in terms of an equivalent amount of the most toxic and most analysed PCDD, *i.e.* 2,3,7,8-TCDD. Thus in the most widely accepted method, 2,3,7,8-TCDD has a TEQ of 1.0 and OCDD, for example, has a TEQ of 0.001.[41]

A number of comprehensive reviews[45-48] have been published dealing with the health effects of PCDD and PCDF, by far the majority of work being on animal tests. Toxicity depends on the number and position of the chlorine substituents,[49] with 2,3,7,8-TCDD being the most toxic. PCDD and PCDF have been shown to cause lethal effects in certain laboratory animals at very low levels;[50] however, it

[45] F. Cattabeni, A. Cavallaro, and G. Galli, 'Dioxin: Toxicological and Chemical Aspects', SP Medical and Scientific Books, London, 1978.

[46] G. Choudhary, L. M. Keith, and C. Rappe, in 'Chlorinated Dioxins and Dibenzofurans in the Total Environment', ed. G. Choudhary, L. M. Keith, and C. Rappe, Butterworth, London, 1983, Section V.

[47] A. W. M. Hay, in 'Chlorinated Dioxins and Related Compounds; Impact on the Environment', ed. O. Hutzinger, R. W. Frei, E. Merian, and F. Pocchiari, Pergamon Press, Oxford, 1982.

[48] S. A. Skene, I. C. Dewhurst, and M. Greenberg, *Hum. Toxicol.*, 1989, **8**, 173.

[49] H. R. Buser and C. Rappe, *Anal. Chem.*, 1984, **56**(3), 442.

[50] P. G. Baker, *Anal. Proc.*, 1981, **18**, 478.

is clear that the toxicological responses are very species-dependent and the extrapolation of animal tests to humans is controversial.

There is less information with respect to the toxic effects on humans and most existing data has been derived from occupational exposure or industrial accident victims. The effects attributed to PCDD and PCDF exposure include a persistent skin acne condition known as chloracne and systemic effects such as digestive disorders and muscle and joint pains, neurological disorders such as headaches and loss of hearing, and psychiatric effects such as depression and sleep disturbance.[47] Also potentially connected to PCDD and PCDF exposure are long term health risks such as chromosome damage, heart attacks, and cancer.[47,51-53] However, Skene *et al.*[48] reported on a number of human accidental and occupational exposures to PCDD and PCDF and their effect on health. The results showed that the human epidemiological studies are difficult to interpret since there have been problems in controlled methodologies, inadequate information on intake and exposure mode and level. In addition, exposures have often been to mixtures of PCDDs and/or PCDFs and also in conjunction with other related and possibly hazardous compounds. Their data suggested that the effects of PCDD and PCDF on humans were inconclusive and required further study. Indeed, an international steering group reporting on the Seveso incident in Italy, where an uncontrolled release of 2,3,7,8-TCDD occurred from a plant manufacturing 2,4,5-trichlorophenol, concluded that no clear-cut adverse health effects attributable to 2,3,7,8-TCDD besides chloracne could be observed.[54] Whilst the evidence for a clear link between exposure to PCDD and PCDF and long-term adverse health effects is inconclusive, the perceived risk is still of some concern to the public.

Table 6 shows the emissions of all the PCDD and PCDF isomers from a range of municipal solid waste incinerators throughout the world.[55-57] The results represent emissions from plants equipped with a range of gas cleaning systems, from older, low efficiency systems, to modern, very sophisticated systems. The results from Hamilton, Canada, and Hampton, USA shown in Table 6 were obtained from old plants with poor combustion control. The Belgian, Netherlands, and German incinerator X are older plants and not considered representative of the modern combustion-controlled plants represented by Prince Edward Island, Canada, and Chicago and Westchester, USA. The lower PCDD and PCDF emissions from the other incinerators are achieved with optimized combustion control. The very low emissions of $<2.0\,ng\,m^{-3}$ are found where efficient and sophisticated gas cleaning systems are incorporated. For example, the Quebec, Canada, incinerator is equipped with water scrubbing, dry lime injection, and fabric filters at controlled temperatures to clean the incinerator gases.[56]

[51] A. Manz, J. Berger, J. H. Dwyer, D. Hesch-Janys, S. Nagel, and H. Walsgott, *Lancet*, 1991, **338**, 959.
[52] R. Saracci, M. Kogevinas, P.-A. Bertazzi, B. H. Mesquita, D. Coggon, L. M. Green, T. Kauppinen, K. A. L'Abbe, M. Littorin, E. Lynge, J. D. Mathews, M. Neuberger, J. Osman, N. Pearce, and R. Winkelmann, *Lancet*, 1991, **338**, 1027.
[53] B. Commoner, K. Shapiro, and T. Webster, *Waste Manage. Res.*, 1987, **5**, 327.
[54] U. G. Ahlborg and K. Victor, *Waste Manage. Res.*, 1987, **5**, 203.
[55] M. J. Suess, *Waste Manage. Res.*, 1987, **5**, 257.
[56] R. Klicius, D. J. Hay, A. Finkelstein, and L. Marentette, *Waste Manage. Res.*, 1987, **5**, 301.
[57] F. Hasselriis, *Waste Manage. Res.*, 1987, **5**, 311.

6 Range of emissions
DD and PCDF from
unicipal Solid Waste
rators (ng m^{-3})[55-57]

Incinerator plant	PCDD	PCDF
Belgium		
Incinerator X	3900	4600
Incinerator Y	840	2900
Canada		
Hamilton, Ont.	1100–7200	3000–10 000
Prince Edward Island	60–190	100–210
Montreal	0.01–0.75	0.02–0.54
Quebec, Ont.	0.4	0.9
Germany		
Neustadt	5	9
Stapelfeld	20–40	90–120
Wurzburg	12–36	10–54
Incinerator X	130–610	300–2400
Netherlands		
Average of 9 plants	1500	1300
USA		
Hampton, VA	500–3800	1600–16 000
Chicago, IL	30–40	170–180
Westchester, NY	15–30	50–80
Marion, OR	0.8–1.5	1.0–2.0

Difficulties arise in comparing plants since they have different capacities, are of different design, and may be operated under different conditions. It has been shown[58] that emissions of PCDDs are very plant-dependent, even when built by the same manufacturer using similar grate types, due to mode of operation, maintenance procedures, age, *etc*.

A number of theories have been proposed for the formation of PCDDs and PCDFs during combustion[59] and their formation route may be a combination of processes, depending on prevailing conditions.

1. PCDD and PCDF occur as trace constituents in the waste and because of their thermal stabilities they survive the combustion process. Waste material has been shown to contain PCDD and PCDF at trace levels;[43] however, mass balances have shown that higher concentrations have been found in the emissions than are found in the input. However, conditions undoubtedly exist for the thermally stable PCDDs and PCDFs to survive the combustion process, particularly at the lower combustion temperatures that prevail in certain zones of some incinerators.
2. PCDDs and PCDFs are produced during the incineration process from precursors such as polychlorinated biphenyls (PCB), chlorinated benzenes, pentachlorophenols, *etc*. The *in situ* synthesis of PCDDs and PCDFs occurs, therefore, via rearrangement, free-radical condensation, dechlorination, and other molecular reactions.[60] These precursors may be present in the waste[43]

[58] M. J. Woodfield, B. Bushby, D. Scott, and K. Webb, in 'Incineration of Municipal Waste', ed. R. B. Dean, Academic Press, London, 1988, p. 332.
[59] J. W. A. Lustenhouwer, K. Olie, and O. Hutzinger, *Chemosphere*, 1980, **9**, 501.

or formed by the combustion of plastics such as PVC.[61,62] However, it has also been shown that there is no direct relationship between incineration of PVC plastic and PCDD and PCDF formation.[14] Several workers have recorded that the presence of PCBs and chlorophenols, *etc.*, in waste can lead to increased emissions of PCDDs and PCDFs.[63,64] The temperature range of formation is from 300 to 800 °C.

3. PCDDs and PCDFs are produced as a result of elementary reactions of the appropriate elements; carbon, hydrogen, oxygen, and chlorine. This reaction is called a *de novo* synthesis of PCDD and PCDF. *De novo* synthesis has been cited to take place in the combustion plasma or in the plume after combustion.[65] PCDD and PCDF also have been shown to form on flyash containing residual carbon collected within a combustion system at temperatures in the region of 300 to 400 °C in the presence of flue gases containing HCl, O_2, and H_2O.[60] It is thought the reaction is catalysed by various metals, metal oxides, silicates, *etc.*, present in the flyash. This theory is borne out by the observation that low levels of PCDDs and PCDFs have been observed in the furnace exit of incinerators but levels 100 times greater were found in the electrostatic precipitator ash of the same plant.[60]

The control of PCDD and PCDF emissions may be approached by; restricting their formation, combustion control, and by clean-up of the flue gases after they have formed. The removal of the chlorine- and HCl-producing plastic components from the waste prior to incineration has been suggested as a mechanism of PCDD and PCDF control. However, results have shown no correlation between PVC plastic in the waste stream or HCl emissions with PCDD and PCDF emissions from incinerators.[14,66,67]

Combustion control has centred on the destruction of PCDDs and PCDFs at high temperatures. Consequently, recommended conditions are temperatures above 1000 °C, residence times of >2 s, and turbulence to ensure good mixing with excess air. Correlations with combustion parameters such as temperature, excess air level (O_2) and CO, and the emission of PCDDs and PCDFs would therefore be expected. The emission of CO from incinerators is used as a measure of efficient combustion, such that minimum CO correlates with efficient combustion. A number of workers have indeed found a correlation between PCDD and PCDF emissions, and furnace temperature, CO, oxygen concentration,

[60] H. Hagenmaier, M. Kraft, R. Haag, and H. Brunner, in 'Energy Recovery Through Waste Combustion', ed. A. Brown, P. Evemy, and G. L. Ferrero, Elsevier Applied Science, Essex, 1988, p. 154.
[61] S. D. Probert, 'Applied Energy: Special Issue on Energy from Refuse', Elsevier Applied Science, Essex, 1987.
[62] S. Marklund, C. Rappe, M. Tysklind, and K. E. Egeback, *Chemosphere*, 1987, **16**, 29.
[63] O. Hutzinger, R. W. Frei, E. Meriam, and F. Pocchiari, 'Chlorinated Dioxins and Related Compounds', Pergamon Press, New York, 1982.
[64] H. Buser, *Chemosphere*, 1979, **8**, 157.
[65] C. Rappe, S. Marklund, A. Bergqvist, and M. Hansson, in 'Chlorinated Dioxins and Dibenzofurans in the Total Environment', ed. G. Choudhary, L. M. Keith, and C. Rappe, Butterworth, London, 1983, p. 99.
[66] S. Nchida and H. Kamo, *Ind. Eng. Chem. Proc. Des. Dev.*, 1983, **22**, 144.
[67] J. R. Visalli, *J. Air Pollut. Control Assoc.*, 1987, **37**, 1451.

and to a lesser extent furnace residence time.[57,68] High combustion efficiencies and high furnace temperature correlating with low flue gas PCDDs and PCDFs emission. However, in contrast, other workers have shown that there is no direct relationship between furnace temperature, CO concentration or combustion efficiency, and PCDD and PCDF emissions.[53,60,69–71] This group of workers suggested that the *de novo* synthesis dominates the formation of PCDDs and PCDFs. PCDDs and PCDFs may be destroyed at the high temperature of the furnace with efficient combustion control, but the overall emissions of PCDDs and PCDFs from the incinerator are not affected by this destruction since formation of these compounds takes place in the cooler parts of the incinerator system, down-stream of the furnace. Commoner *et al.*[53] showed that PCDD and PCDF emissions from an incinerator furnace outlet were negligible, but much larger concentrations were found in the cooler parts of the incinerator system prior to the stack due to *de novo* synthesis in the heat exchangers (Figure 3). Figure 3 also shows that the incinerator is effective in destroying the PCDDs where a high input of PCDDs is reduced to negligible concentrations, but again *de novo* synthesis causes a dramatic formation of PCDDs in the heat exchangers.

The furnace temperature and operating conditions are however important in reducing flue gas PCDD and PCDF emissions. Efficient furnace combustion conditions will minimize the production of the products of incomplete combustion and particulate carbon which then minimizes the extent of the *de novo* reaction on the flyash surface to form PCDD and PCDF. In addition, if the gas temperature inlet to the ash collection system can be reduced to below 200 °C this will reduce the *de novo* reaction which is most significant at between 250 and 450 °C.

Post-combustion control of PCDDs and PCDFs has centred on the efficient collection of particulates since they are shown to be found mostly on flyash, either adsorbed or formed *in situ*; they also exist at lower levels in the gas phase. Efficient collection of particulates utilizes electrostatic precipitators and fabric filters with the latter showing better retention for PCDDs and PCDFs.[15] Klicius *et al.*[56] have shown that gas cleaning by dry sorption or wet scrubbing and subsequent removal of dust by fabric filters at temperatures below 150 °C results in a marked reduction of PCDDs in stack emissions. Wet/dry scrubbers, with lime slurry as the active scrubbing agent, have also been shown to be effective in the removal of PCDDs and PCDFs.[72] The addition of small quantities of activated carbon to the alkaline adsorbent have shown very high removal rates for PCDD and PCDF. Other workers[73] have suggested that the activated carbon is utilized in a series of coke beds for cleaning the flue gases.

The emission of PCDDs and PCDFs from waste incinerators can be compared with other combustion sources. Difficulties arise in comparisons of different types of plant, however, because of variations in combustion controls, efficiency, plant

[68] J. G. T. Bergstrom and K. Warman, *Waste Manage. Res.*, 1987, **5**, 395.
[69] H. Hagenmaier, H. Brunner, R. Haag, M. Kraft, and K. Lutzke, *Waste Manage. Res.*, 1987, **5**, 239.
[70] H. Hagenmaier, M. Kraft, M. Brunner, and R. Haag, *Environ. Sci. Technol.*, 1987, **21**, 1080.
[71] H. Vogg, M. Metzger, and L. Stieglitz, *Waste Manage. Res.*, 1987, **5**, 285.
[72] K. K. Neilsen, J. T. Moeller, and S. Rasmussen, *Chemosphere*, 1986, **15**, 1247.
[73] A. G. Buekens and F. De Geyter, 'Municipal Waste Combustion Developments in Europe', Municipal Waste Combustion Conference of the Air and Waste Management Association, 15–19 April, Florida, USA, 1991.

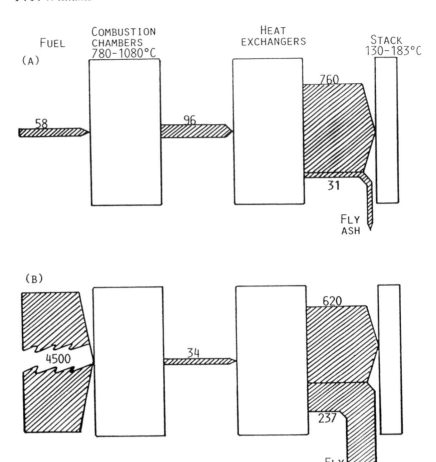

Figure 3 Diagrammatic presentation of the mass balance of dioxins and furans for a Canadian incinerator (1985); (a) furans, (b) dioxins. Mass balance data are in $\mu g\,h^{-1}$. (Figure (ref. 53) reproduced with permission from *Waste Manage. Res.*, 1987, **5**, 327.)

size, fuel type, and so on. There are conflicting and limited data on the emission of PCDDs and PCDFs from coal-fired power stations: some studies have shown no PCDDs or PCDFs in samples of flyash, whereas others have shown trace levels of PCDDs and PCDFs.[41] Kimble and Gross[74] reported emission levels of less than 0.005 ng TCDD kg^{-1} of coal for a coal-fired power station in the USA. Similarly, for industrial coal burning there are limited emission data for PCDDs and PCDFs. However, emission levels from two tests in the UK have shown 14 ng and 18 ng total TCDD kg^{-1} of coal combusted, and 23 and 162 ng total TCDD kg^{-1} of coal combusted.[41] The levels of PCDDs and PCDFs from coal combustion are lower than from waste incinerators; however, the mass of coal combusted relative to waste is much higher, and consequently the overall burden to the atmosphere from this source may be higher than from incineration.

The combustion of hospital waste is a further potential source of PCDDs and PCDFs. The limited data available have shown that emission levels are similar to

[74] B. J. Kimble and M. L. Gross, *Science*, 1980, **207**, 59.

those from the incineration of municipal waste: for example, 9 ng TCDD kg^{-1} of waste combusted from one UK plant, and 290 ng TCDD kg^{-1} of waste combusted from a USA hospital waste incinerator.[41] PCDDs and PCDFs are produced at low levels from certain motor vehicle exhausts using leaded gasoline; the scavenger dichloroethane acts as a precursor for the PCDD and PCDF.[62] PCDDs and PCDFs have also been detected in used lubricating oil derived from chlorinated additives in the oil or gasoline,[75] and have also been detected in the emissions from wood-burning stoves and from cigarettes.[1]

It is interesting to note that Hagenmaier *et al.*[69] have suggested that the occurrence of PCDDs and PCDFs in such widespread environments as sediments, soil, sewage sludge, *etc.*, indicates that their source is not mainly from the incineration of municipal waste but must be from elsewhere, such as the production of pentachlorophenol used as a wood preservative. PCDDs and PCDFs are found as impurities in pentachlorophenol and Hagenmaier *et al.*[69] have calculated that the input of PCDDs and PCDFs to the environment from this source far exceeds the input from combustion sources.

3 Contaminated Waste Water

Water pollution from incinerators is not generally regarded as an important problem because of the limited amount of waste water generated, *i.e.* of the order of 2.5 m^3 tonne^{-1} of municipal waste incinerated. However, the waste water from municipal waste incinerator plants has been shown to be contaminated with heavy metals and inorganic salts and to have high temperatures and high acidities or alkalinities.[76]

The main sources of waste water from incinerators are flue gas treatment, as flue gas scrubber water and alkaline scrubbing of the gases to remove acid gases, and the quenching of incinerator ash. Other minor sources include, for example, scrubber water pre-treatment and the purification of boiler feed-water where a boiler plant is installed. Where incinerators have no form of heat recovery the gases from the furnace are cooled by water injection. The water is evaporated completely and passes to the gas clean-up system.

Reimann[76] has analysed the raw flue-gas scrubber water before any neutralization, from a German incinerator. He showed (Table 7) that pH levels were very low, between 0.5 and 1.3, and heavy metal levels (in particular, nickel, zinc, cadmium, lead, and mercury) exceeded permitted disposal conditions. The heavy metals were neutralized in the flue-gas waste water via precipitation with Ca(OH)$_2$, which causes the major part of the pollutants to be precipitated as hydroxide sludges. However, more than 60% of the mercury and other heavy metals remain in the waste water and the addition of TMT 15 (trimercapto-*s*-triazine) is required, which precipitates the mercury and heavy metals down to levels well below the permissible limits.[76]

The scrubbing of the flue-gases with alkaline NaOH solutions to remove acidic gases also generates waste water. The scrubber water produces sulfite from the dissolution of sulfur dioxide from the flue gases, which is then oxidized to sulfate

[75] K. Ballschmiter, H. Buchert, R. Niemczyk, A. Munder, and M. Swerev, *Chemosphere*, 1986, **15**, 901.

[76] D. O. Reimann, *Waste Manage. Res.*, 1987, **5**, 147.

Table 7 Concentration of
pollutants in incinerator
waste water (mg l^{-1})[77]

Pollutant	Flue gas scrubber water	Bottom ash waste water
pH	0.95	8.8
Cl	12 900	1540
SO$_4$	502	590
F	52	1.7
Cr	0.69	0.10
Cu	1.28	0.26
Ni	3.7	0.25
Zn	14.1	1.8
Cd	0.46	0.15
Pb	6.8	0.80
Hg	6.6	0.038

by oxygen in the flue gas. The sulfate is precipitated as $CaSO_4$ due to the presence of calcium ions either from the first flue gas scrubber or added as $Ca(OH)_2$. The waste water from the alkaline scrubber will have a pH of over 11 and therefore requires acidification to neutralize it. The waste water may also contain heavy metals and consequently may require neutralization with TMT 15, for example, to remove the heavy metals.

The bottom-ash from the incinerator grate is removed in a unit which serves to cool the ash and also maintain a partial vacuum in the incinerator chamber. Approximately 0.3 to 0.8 m^3 of water is used per tonne of waste.[76] Table 7 shows the contamination of bottom-ash wash-water analysed after a two-hour sedimentation period to allow settling of the solid particulate material. The waste water is alkaline, with a mean pH of 8.8. The data for heavy metal pollution showed that mean figures over a three year period were well below the permitted disposal levels; however, occasionally the concentrations exceeded the permitted levels.

Whilst there is most concern over the presence of heavy metals in waste water, the presence of organic pollutants such as polycyclic aromatic compounds (PAC) and dioxins and furans should also be suspected. Reimann[76] reports the presence of polycyclic aromatic compounds at very low levels, of the order of a total PAC concentration of 0.05 μg l^{-1} from the untreated raw waste water from a German incinerator. The concentration of PAC remained unchanged after neutralization treatment with additives. Davies et al.[33] reports individual PAC in the wastewater output as bottom-ash wash-water and flue-gas coolant water from a UK incinerator. The reported concentrations of fluoranthene, pyrene, perylene, benzo(a)pyrene, and benzo(e)pyrene in the output waste water were 0.62, 0.54, 0.13, 0.14, and 0.14 μg l^{-1}, respectively. Dioxins and furans could not be detected by Ozvacic et al.[77] in the water effluent from the ash handling system of a Canadian incinerator; however, they were present in high concentration,

[77] V. Ozvacic, G. Wong, H. Tosine, R. E. Clement, and J. Osborne, J. Air Pollut. Control Assoc., 1985, 35, 849.

Pollutant	Bottom-ash[a] USA	Flyash[a] USA	Bottom-ash[b] Germany	Flyash[b] Germany
Cd	41	64	3.8	—
Cr	520	1160	655	450
Cu	450	510	1520	1050
Hg	0.4	0.9	0.7	8
Mn	3100	1500	—	—
Ni	210	1800	260	145
Pb	1700	7200	1010	4900
Zn	5500	10 000	4570	16 600
Cl	—	—	1970	38 000
F	—	—	215	105

[a]Ref. 14.
[b]Ref. 12.

1565 ng l^{-1}, in the suspended particulate. Bumb *et al.*[78] also analysed scrubber water from a rotary kiln incinerator and reported that 99.7% of the PCDDs in the water were associated with the suspended particles which were filtered out of the water and only very low levels, of the order of 0.01 p.p.b., were present as dissolved species in the water itself.

4 Contaminated Ash

If the incinerator is operating correctly, the residue or ash should be completely burnt out and biologically sterile. Bottom-ash from the furnace grate represents the bulk (75–90%) of total ash and is composed mainly of mineral oxides. Its heavy metal content is generally lower than 1.5% but is highly variable. The ash is usually disposed of to land.

Table 8 shows concentrations of heavy metals in bottom-ash and flyash from USA and German incinerators.[12,14] The concentrations of heavy metals clearly are high.

The pollutants present in the ash residues from incineration become of more significance when they are placed in landfill sites where leaching of the pollutants may be a source of groundwater contamination. Sawell *et al.*[79] have shown that metal leachability is dependent upon metal speciation, the pH of the leaching medium, and ash particle size. It has been shown [14] that flyash is more readily leached than the clinker fraction, with 2–5 g water-soluble matter kg^{-1} of dry clinker and 100 g water-soluble matter kg^{-1} of dry flyash. The leachability of metals adsorbed on flyash particles is enhanced, since the heavy metals largely occur in the smallest size fraction of less than 10 μm and are concentrated at or near the surface of the particles.[10,18,19] The small particle size increases the available surface area exposed to the leaching fluid. In addition, the high chlorine

[78] R. R. Bumb, W. B. Crummett, S. S. Cutie, J. R. Gledhill, R. H. Hummel, R. O. Kagel, L. L. Lamparski, E. V. Luoma, D. L. Miller, T. J. Nestrick, L. A. Shadoff, R. H. Stehl, and J. S. Woods, *Science*, 1980, **210**, 385.

[79] S. E. Sawell, T. R. Bridle, and T. W. Constable, *Waste Manage. Res.*, 1988, **6**, 227.

Table 9 Leachate from solid incinerator residues $(mg\,l^{-1})$[14]

Pollutant	Concentration
Cl	6500–20 000
SO_4	70–1300
F	0.4–1.7
Pb	0.1–0.9
Cr	0.05–0.10
Cu	0.05–0.30
Zn	0.05–0.30
Cd	0.02–0.15
Ni	0.05–0.60

content of the waste results in the majority of the metal species being present as the metal chlorides which are generally more soluble in water than other species.[9,19] It has been shown that up to 32.5% of the available zinc, 1.75% of lead, 5.7% of manganese, and 94% of the available cadmium can be leached from flyash.[14] Sawell et al.[79] also have confirmed the high solubilities of cadmium, zinc, lead, and copper leached from municipal waste-incinerator ash. Most leachability tests have been undertaken under laboratory conditions using distilled water; however, leaching of the metals from the ash is greatly increased if acidic solutions, which attempt to simulate acid rain conditions, are used. Water in contact with flyash produces alkaline solutions rather than acidic ones; Sawell et al.[79] and Denison and Silbergeld[19] have shown that copper, lead, zinc, and cadmium show increased solubilities at high alkalinities; that is, they are amphoteric in nature, showing significant solubilities at both low and high pH values. There are less data on the analysis of leachate from real municipal waste ash disposal sites.

Buekens and Schoeter[14] (Table 9) showed that concentrations of leachate from a combined bottom-ash and flyash disposal site yielded concentrations of heavy metals which were of similar concentration to laboratory leaching tests. Carlsson[80] has shown concentrations of heavy metal leachate for solidified municipal solid waste ash from a landfill site and showed that solidification of the ash produced lower leachability than unprocessed ash.

PACs have also been detected in the ash derived from incinerators.[14,33,81] Table 10 shows PAC found in the clinker fraction and flyash of various incinerators. The clinker fraction showed higher concentrations of PAC than the flyash for the Dutch incinerator. The PACs found have been shown to include mutagenic and/or carcinogenic compounds and to occur at significant concentrations; for example, benzo(a)pyrene, benzo(a)anthracene, benzofluoranthenes, etc. Davies et al.[33] suggest that leaching of PAC from ash disposal in landfill sites may result in contamination of ground water.

The presence of PCDDs and PCDFs has been shown in flyashes derived from the incineration of solid waste. Table 11 shows total PCDDs and PCDFs and

[80] B. Carlsson, 'Solidification of Flue Gas Cleaning Products', Proceedings of the Conference on Control of Incinerator Pollution, June, Birmingham, UK, 1991.

[81] G. A. Eicemann, R. E. Clement, and F. W. Karasek, *Anal. Chem.*, 1981, **53**, 955.

0 Concentration of in incinerator solid ues $(\mu g\, g^{-1})$[14,33,82]

Pollutant	Holland[a] Clinker	Holland[a] Flyash	Canada[b] Flyash	UK[c] Flyash
Fluorene	145	40	<0.5–64	—
Phenanthrene	245	120	—	—
Fluoranthene	350	70	0.5–440	58
Pyrene	470	155	0.5–120	49
Benz(a)anthracene	105	25	—	171
Chrysene	180	45	—	
Benzofluoranthenes	170	36	—	—
Benzo(a)pyrene	60	10	0–14*	147*

*Benzo(a)pyrene + Benzo(e)pyrene.
[a]Ref. 14.
[b]Ref. 82.
[c]Ref. 33.

1 Concentration of CDD and PCDF in rator solid residues $(ng\, g^{-1})$[69,82,83]

PCDD/PCDF	Germany[a]	Canada[b]	Netherlands[c]
TCDD	11	13	93
PeCDD	34	23	254
HxCDD	50	26	604
HpCDD	57	15	760
OCDD	65	6	345
Total PCDD	210	83	2056
TCDF	72	—	173
PeCDF	95	—	312
HxCDF	82	—	459
HpCDF	56	—	314
OCDF	13	—	51
Total PCDF	275	—	1309

[a]Ref. 69.
[b]Ref. 82.
[c]Ref. 83.

certain isomers found in flyashes from German, Canadian, and Dutch incinerators.[69,81,82] The German data represented the average of 52 flyash samples from 10 incinerators and the Canadian data an average of 8 samples. The significant concentrations of PCDDs and PCDFs found in flyash samples illustrates the *de novo* synthesis route to their formation, a route catalysed by the flyash itself. Ozvacic *et al.*[77] have shown for a Canadian incinerator that furnace

[82] K. Olie, J. W. A. Lustenhouwer, and O. Hutzinger, in 'Chlorinated Dioxins and Related Compounds; Impact on the Environment', ed. O. Hutzinger, R. W. Frei, E. Merian, and F. Pocchiari, Pergamon Press, Oxford, 1982, p. 227.

bottom-ash contains negligible concentration of PCDDs and PCDFs but electrostatic precipitator ash contained high concentrations. This has also been confirmed in later work by Shaub,[83] suggesting that optimum *de novo* synthesis conditions occur in the flyash collection system but that bottom-ashes are quickly quenched before significant PCDD and PCDF production can take place.

[83] W. M. Shaub, 'An Overview of What is Known About Dioxin and Furan Formation, Destruction, and Control During Incineration of MSW', Coalition on Resource, Recovery and the Environment (CORRE), USA EPA MSW Technology Conference, San Diego, CA, Jan.–Feb., 1989.

Recovering Energy from Waste: Emissions and Their Control

G. W. RAE

1 Introduction

Although reliable data are difficult to find, it is estimated by the Department of the Environment[1] that some 516 million tonnes of waste arise annually in the UK. Of this some 137 million tonnes is classified as household, commercial, and industrial waste. These are the controlled wastes for which private sector companies are responsible at a range of waste management facilities. The proper disposal of these wastes is essential if public health and the environment are to be protected. A range of disposal facilities are provided to deal with these wastes. But by far the most significant disposal route is landfilling. Some 85 to 90% of all controlled waste go untreated to landfill, with landfill also acting as the final repository for the residues from such waste treatment and waste reduction processes as are available. Currently there are some 6000 licensed landfill sites in the UK.

The only serious, though limited, alternative to landfill for most controlled wastes is incineration. Currently some 30 incinerators, all built by local authorities between 1968 and 1976, are operational. All of these are of the mass burn type, accepting waste without pre-processing, and most operate with a throughput of between 6 to 10 tonnes h^{-1}. Only five of these plants recover energy from the wastes. Collectively these incinerators deal with some 7% of UK household and commercial waste. There is, however, little scope for increasing the throughput at these existing plants. The technologies employed are outmoded, particularly in terms of emission control, and without extensive retrofitting none of these plants will meet the standards set out in the EC Directive on the Incineration of Municipal Waste.[2] This directive comes into force in 1996 and at that time it is expected that all but one or two of these existing incinerators will close.

This contrasts with the situation in Europe where incineration makes a substantial contribution towards the disposal of household and commercial

---

[1] Department of the Environment, 'A Review of Options—A Memorandum Providing Guidance on the Options Available for Waste Treatment and Disposal', Waste Management Paper No. 1, HMSO, London, 1992.

[2] Directive on the Reduction of Air Pollution from Existing Municipal Waste Incineration Plant, 89/429/EEC, OJ No. L203,15.7.89, p. 50.

waste. For example, incineration accounts for 60% of domestic waste disposal in Denmark; 35% in France; 35% in Germany; and 30% in the Netherlands, and the figure is increasing.[1] These plants invariably are equipped with some form of energy recovery and have emission control systems that routinely exceed the current EC standards for incineration. However, due to tighter environmental standards as required by the Environmental Protection Act, 1990,[3] and Government initiatives concerning the use of renewable energy, the picture within the UK is changing. A number of waste management companies have signalled a willingness to invest in the incineration of domestic and commercial waste in order to recover its energy value. The schemes proposed rely heavily on the very best continental technologies, particularly in terms of efficiency of energy recovery and emission control.

As a developer of waste-to-energy plant within the UK, Cory Environmental is currently involved in schemes which collectively would deal with some 2.5 million tonnes of waste annually. An essential requirement in taking these forward is a commitment to very high standards of emission control. This paper, by reference to the proposed waste-to-energy plant at Belvedere in the London Borough of Bexley, examines the control technologies that are deployed to ensure that such plants are environmentally acceptable even when located, as they have to be for logistical reasons, close to or at centres of urbanization.

2 The Belvedere Project

Currently much of central and east London's waste goes to landfills in Kent and Essex (largely rural counties adjacent to greater London), with some travelling further to other counties. A survey by the London and South East Regional Conference (SERPLAN) in 1991[4] showed the volume of void space permitted or licensed for waste disposal in South East England was 222 million cubic metres. Average solid waste arisings in the SERPLAN surveys since 1987 have been at a rate of 31.6 million tonnes annum^{-1}, calculated as consuming 30.4 million m^3 of void. This clearly varies with economic activity and population changes, and the 1991 survey showed that 28 million tonnes of waste was disposed of in that year.

At the levels of disposal of the recent past, the 222 million m^3 of permitted void will last for only some 7 years. Even if sites that could be used as landfill, but do not have the necessary permissions, are taken into account the void will only last for 8 years (*i.e.* run out in 1999). Nor is there any realistic prospect of substantial quantities of new landfill capacity being released in the future. Planning and environmental policies, at both national and regional level, are serving to severely restrict the availability of new landfill.

It is against this background that Cory Environmental has sought permission to construct and operate a waste-to-energy plant at Belvedere in the London Borough of Bexley. The location of the proposed plant is shown in Figure 1 whilst an artist's impression of the plant is in Figure 2. A diagrammatic cross-section of the plant is given in Figure 3. The numbers in parenthesis in the description of the

[3] The Environmental Protection Act, 1990, Chapter 43, HMSO, London, 1990.

[4] The London and South East Planning Conference, 'Waste Disposal in South East England: Results of the 1991 Waste Monitoring Survey', RPC 2090, London, 1992.

Figure 1 Location of proposed Belvedere refuse-to-energy plant

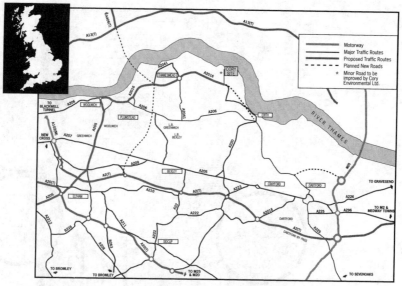

LOCATION

Figure 2 Artist's impression of the plant

ARTIST'S IMPRESSION

plant's operation below refer to this figure. The plant will produce energy from waste by converting as much as possible of the energy in domestic and commercial waste into electricity. The plant consists of four streams each of a design capacity of 38.5 tonnes h^{-1} which together will handle 1.2 million tonnes of municipal waste $annum^{-1}$. Some 103 MW of electricity will be produced for export to the National Distribution Network. The full output of the station was

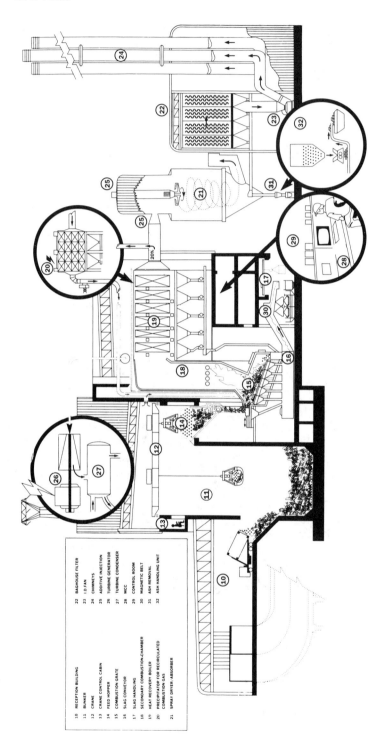

Figure 3 Cross-section of the plant.

10	RECEPTION BUILDING	22	BAGHOUSE FILTER
11	BUNKER	23	I.D FAN
12	CRANE	24	CHIMNEYS
13	CRANE CONTROL CABIN	25	ADDITIVE INJECTION
14	FEED HOPPER	26	TURBINE GENERATOR
15	COMBUSTION GRATE	27	TURBINE CONDENSER
16	SLAG CONVEYOR	28	MCC
17	SLAG HANDLING	29	CONTROL ROOM
18	SECONDARY COMBUSTION CHAMBER	30	MAGNETIC BELT
19	HEAT RECOVERY BOILER	31	ASH REMOVAL
20	PRECIPITATOR FOR RECIRCULATED COMBUSTION GAS	32	ASH HANDLING UNIT
21	SPRAY DRYER / ABSORBER		

56

included in the latest tranche of the Non Fossil Fuel Obligation which runs until 1998. The Belvedere plant represents 22.5% of this order and 39.4% of the waste band. The Non Fossil Fuel Obligation is the Government's chosen fiscal instrument to encourage the use of renewable energy sources. It provides initial financial support thereby allowing waste-to-energy to compete with other forms of waste disposal in the right circumstances.

Three types of waste delivery are proposed. First waste in 20 ft ISO containers (average 12.5 tonnes of waste per container) will be delivered by barge. As much waste as possible, probably 850 ktonnes annum^{-1}, and possibly as much as 1 million tonnes annum^{-1} will be delivered by this system. Secondly, between 200 and 400 ktonnes annum^{-1} will be delivered by road either in 20 ft ISO containers or in bulk vehicles. Finally provision has been made for refuse collection vehicles operating in the London Borough of Bexley to deliver directly to the site. Ash and other residuals will be placed in sealed 20 ft ISO containers and barged to a riverside landfill site for final disposal.

Turning to the operation of the plant (Figure 3), all waste is drawn from the storage bunker [11 in Figure 3] by one of the two grab cranes [12] and fed into one of the four feed hoppers serving the four lines of the plant. The wastes are pushed by hydraulic ram at the appropriate rate onto the grate [15] where the wastes are burned under controlled conditions to release the maximum heat value. Solid residues from the burning process, known as clinker, fall into a sealed water bath [16] where they are cooled before being conveyed, via the ferrous metal recovery plant [30], to the ash boxes.

The hot gases given off by the burning wastes pass via the secondary combustion chamber [18], where in exceptional circumstances they may be further heated to ensure compliance with the design parameters, to the boiler [19]. This highly efficient boiler converts the energy of the hot gases into steam through a series of heat exchangers and superheaters. The steam is led off from the boiler to the turbine house [26] where it is used to power conventional turbines to produce electricity. Ammonia is injected into the boiler to ensure low emissions of oxides of nitrogen, as discussed in more detail later. The turbines are cooled by water drawn from the River Thames, augmented by auxiliary cooling during periods of high river temperatures.

The exhaust gases leaving the boiler carry dust and metals and have to be cleaned before being emitted to atmosphere. This is described in detail later as it forms the main focus of this paper. In brief, some 17% of the exhaust gases are recirculated through an electrostatic precipitator [20] and reinjected as secondary air into the furnace. The reinjection of this hot secondary air increases the thermal efficiency of the process and minimizes the creation of pollutants.

The remaining gases pass through a spray drier/absorber [21] where water, mixed with lime slurry, is injected into the gas stream. In addition, activated carbon is also injected at this stage to ensure the almost complete removal of dioxins and heavy metals such as cadmium and mercury from the gases. Finally, the gases pass through a baghouse filter [22] where the majority of the remaining entrained particulates are removed.

The residues from the electrostatic precipitator are conveyed to the ash handling system [32] where they are mixed with the ash and inerts from the furnace. The residues from the gas cleaning system are first pressed into briquettes to minimize dust and ease handling. The use of the ash and residues from the furnace, though not the gas cleaning system, is currently being investigated as sub-base for construction projects. This is a practice adopted widely in continental Europe.

3 The Regulatory Environment

Before considering the emission control system of a waste-to-energy plant in more detail, it is necessary to divert slightly and consider the regulatory environment within which such plants operate. It is this environment that sets the framework for the standards the plant has to meet. Before a waste-to-energy plant can be operated in the UK consent has to be sought from the Department of Trade and Industry to construct and operate a generating station under Section 36 of the Electricity Act, 1989.[5] In this respect waste-to-energy plants are treated no differently from conventional power stations. (This only applies to plants producing more than 50 MW. Plants with ratings below this need apply only for planning permission in the usual way.)

The system under the Electricity Act is analogous to the system under the Town and Country Planning Acts. The local authority in whose area the plant is to be sited is consulted as are other statutory authorities such as the National Rivers Authority, English Nature, and the like. In granting permission the Department can impose conditions relating to the development and operation of the plant. Though it is Government's stated intention, as set out in Planning Policy Guidance Note 1,[6] that the planning system should not duplicate the more specific controls of other systems, notably that set up by the Environmental Protection Act, 1990,[3] it is noteworthy that the whole of the Public Inquiry into the Belvedere plant was concerned with public health and environmental issues and not with electricity generation. Accordingly, from the process of obtaining permission to construct and operate a waste-to-energy plant, the operator must anticipate some conditions governing the environmental impact of the plant. However, these will usually relate to matters such as noise and traffic and should not deal with standards relating to, for example, emissions and releases from the plant.

The specific control system regulating the environmental impact of waste-to-energy plants is the system of Integrated Pollution Control (IPC) set up by Part I of the Environmental Protection Act, 1990 and administered by HM Inspectorate of Pollution. IPC seeks to regulate emissions to the environment as a whole, rather than to regulate the three environmental media separately. In the past emissions have been controlled by three separate regulatory bodies and limits set by one could often conflict with those set by another. The performance standards for waste-to-energy plants are set out in Process Guidance Note IPR

---

[5] Electricity Act, 1989, Chapter 29, HMSO, London, 1989.
[6] Department of the Environment, 'Planning Policy Guidance Note, General Policy, and Principles (PPG 1)', HMSO, London, 1992.

Table 1 Composition of guaranteed emissions from proposed Belvedere with current national and EC standards

Substance mg $(Nm^3)^{-1}$	Belvedere guarantee	HMIP Standard	Standard	EC German (BIm SchV-17)
HCl	30	30	50	10
HF	1	2	2	1
SO_2	70	300	300	50
NO_x	200	350	—	200
Particulates	10	30	30	10
Carbon (excl. particulates)	20	20	20	10
CO	50	100	100	50
T(4)CDD*/ng $(Nm^3)^{-1}$ T(4)CDF*/ng $(Nm^3)^{-1}$	0.1	1	—	0.1
Hg	0.1	0.1	0.2	0.1
Cd and Ti	0.1	0.1	—	—
As and Ni	1.0	1.0	1.0	0.5
Pb, Cr, Mn, and Cu	1.0	—	5.0	—

Nm^3 refers to standard conditions 273 K, 101.3 KPa, 11% O_2 dry gas.
*T(4)CDD = tetrachlorodibenzo-*p*-dioxins (sum of congeners)
 T(4)CDF = tetrachlorodibenzofurans (sum of congeners)

5/3.[7] This Note, though specifically aimed at municipal waste incinerators with a capacity of 1 tonne or more h^{-1} also applies to waste-to-energy plants. It is the aim of this guidance to ensure that plants use best available techniques not entailing excessive cost (BATNEEC) to prevent the release of prescribed substances into the environment and, where that is not practicable to reduce or render harmless unavoidable releases.

Limits for the release of substances into the atmosphere are given in the Note and these are set out in Table 1. It is interesting to contrast these with the current EC standard and the latest German standard, both of which are also set out in Table 1. Although in law the HMIP standard prevails, it is clear that the latest German standard must, in fact, represent the best available technique (BAT), though not necessarily BATNEEC. It is, however, clearly indicative of the direction in which standards are heading in the short- to medium-term.

4 Emissions from Burning Waste

Waste from households, shops, and commercial and industrial premises is an exceedingly heterogeneous material whose composition varies on a day-to-day basis. Yet a waste-to-energy plant has to be built against some standard for the waste it is to burn and the raw gases it is to treat. The key parameters are the net calorific value of the waste and the composition of the raw gases after combustion. In order to understand the make up of the wastes the Belvedere plant is likely to handle, a wide ranging sampling exercise of London's wastes was undertaken.

[7] HM Inspectorate of Pollution, Chief Inspector's Guidance to Inspectors, Environmental Protection Act 1990, Process Guidance Note IPR 5/3, 'Waste Disposal and Recycling Municipal Waste Incineration', HMSO, London, 1992.

By content	%
Total moisture	25.7
Ash	25.0
Carbon	31.8
Hydrogen	4.1
Oxygen	12.4
Nitrogen	0.4
Sulfur	0.6
	100.0

By component	% (by weight)
Paper	30.6
Plastics	8.4
Textiles	1.9
Misc. combustibles	5.5
Misc. non-combustibles	1.7
Glass	9.5
Putrescibles	28.0
Ferrous metal	7.0
Non-ferrous metal	0.6
<10 mm fines	6.8
	100.0

Table 2 'Average' composition of London's waste

Average lower calorific heating value $10\,260\,kJ\,kg^{-1}$

This allowed specification of an average waste against which to design the plant. This average waste is set out in Table 2. From this table it is clear that increased recycling of metals and glass will have a strong beneficial effect in increasing the calorific value of the waste. This echoes the experience elsewhere in Europe where increased recycling has led to a substantially increased calorific value of the waste.

The other point to note is that as a fuel waste is not ideal. It has no more, at best, than half the calorific value of coal. The wide range of components that make up waste, notably the plastics, also mean that the raw gases given off during combustion are heavily contaminated with pollutants. This is best seen in Table 3 which gives the anticipated raw gas composition for the Belvedere plant. These gases need to be treated and conditioned before they can be discharged to atmosphere. To ensure that these gases pose no risk to human health or the environment, the Belvedere plant will use a combination of well proven treatment technologies allied with evolving chemical control systems to ensure that its emissions to atmosphere are always well within current standards.

5 Controlling Emissions to Atmosphere

Referring back to Figure 3, the key technologies employed to treat gas are, for each of the four lines, an electrostatic precipitator, a semi-dry spray absorber and a baghouse filter. How these combine to treat the gas is now described.

Raw gas prior to cleaning

% (vol.) wet	
CO_2	10.8
H_2O	16.2
N_2	67.0
O_2	6.0
	100.0

Pollutants in gas prior to treatment mg $(Nm^3)^{-1}$	
Dust	2500
HCl	1000
HF	10
SO_2	300
Hg	0.8
NO_x	350
CO	50
PCDD, PCDF	5.0

(Nm^3 refer to cubic metres under standard conditions: 273 K, 101.3 kPa, 11% O_2, dry gas.)

NO_x Reduction

Oxides of nitrogen (NO_x) can be formed either by the oxygenation of nitrogen in the waste (known as fuel NO_x) or by high temperature fixation of nitrogen in the combustion air. The formation of fuel NO_x is determined by the nitrogen content of the waste, total excess air rates, and the relative distribution of primary and secondary combustion air. The formation of thermal NO_x on the other hand depends on oxygen availability and the temperature, pressure, and residence time of the gas in the combustion unit. The relative percentages of fuel and thermal NO_x are determined by the characteristics of the waste and by the design of the combustion unit. But for waste-to-energy plants such as Belvedere up to 80% of NO_x in the raw gases may be fuel NO_x. As it is not possible to control or reduce the nitrogen content of the waste, control of NO_x is achieved by having favourable air mixing and distribution conditions.

At Belvedere flue gas recirculation will be used to modify the combustion conditions. Some 17% of the flue gas will be recirculated, via the electrostatic precipitator, to the combustion chamber. (The electrostatic precipitator is used to dedust the gas down to 100 mg $(Nm^3)^{-1}$ in order to protect the combustion air fans against erosion.) This will allow combustion to take place at lower peak temperatures and in conditions of reduced oxygen availability. The former reduces the amount of thermal NO_x while the latter limits the amount of oxygen available to react with nitrogen. In addition, ammonia will be injected into the boiler to further reduce the amount of NO_x in the gas stream. The net effect of this treatment system will be a reduction of NO_x of approximately 20% and an improvement in boiler efficiency of between 2 to 3%. It will result in the gases being exhausted to atmosphere always having a NO_x content below

61

200 mm $(Nm^3)^{-1}$. This is well within the requirements of Her Majesty's Inspectorate of Pollution and corresponds with the latest German standard.

Dioxins and Furans

Measures will be taken both during and after combustion to prevent the formation, reformation, and release of PCDD's and PCDF's. Combustion conditions have been optimized so as to achieve maximum burnout of the waste. In this way only a minimum amount of products of incomplete combustion (PICs), which are the precursors of dioxins/furans, are formed. Normal operating combustion conditions at the Belvedere plant will provide a temperature of 1000–1100 °C, a residence time of greater than 2 s, and good mixing. All of these have been shown to be effective in preventing the formation of dioxins and furans. Removal of PCDD's and PCDF's from the gas stream will be accomplished by scrubbing the gases and by the injection of activated carbon into the gas stream. The system to be employed consists of a spray absorber followed by a baghouse filter. The spray absorber consists of a cylindrical vessel, 11 m in diameter and 11 m in height, with a conical bottom section. A mixture of hydrate of lime and activated carbon is injected via a rotating disk into the cylinder where it forms a fine mist which mixes intimately with the flue gas. Whilst the hydrate of lime is injected primarily to control the acid gases, activated carbon is injected primarily to control dioxins and furans and heavy metals such as cadmium and mercury.

Dioxins have been shown to cling to particulate matter and studies by Lurgi, who are supplying the gas cleaning train for the Belvedere plant, have shown activated carbon to be particularly efficient at scavenging dioxins and furans from the gas stream. As a side effect, the injection of activated carbon is also helpful in controlling the emissions of mercury and cadmium. Mercury and cadmium are relatively volatile compared with other heavy metals and hence have a tendency to pass through conventional gas cleaning equipment in the vapour phase.

Whilst some of the particulate matter will fall out in the spray absorber, most will pass through. Indeed, due to the injection of lime and activated carbon the particulate burden of the flue gas is likely to be higher exiting the spray absorber than it was on entry. This particulate load has to be removed, and at the Belvedere plant baghouse filters will be employed. Each filter, which consists of 1792 bags giving a surface area of 4900 m^2, has been designed to have a collection efficiency of 99.8%. It will also have an efficiency of greater than 95% in removing particulates of less than 0.2 μm. This combination of activated carbon injection and baghouse filter allows the process supplier, Lurgi, to guarantee emissions of 0.1 ng $(Nm^3)^{-1}$ for dioxins and furans. In addition they are prepared to guarantee 0.1 mg $(Nm^3)^{-1}$ for mercury and 0.1 mg $(Nm^3)^{-1}$ for cadmium and thallium. These are significantly better than current UK or EC standards.

Acidic Gases

The main acidic gases arising form the burning of household and commercial wastes are hydrochloric acid, hydrofluoric acid, and sulfuric acid. A number of technologies are available for their removal; notably spray towers, packed bed scrubbers, semi-dry scrubbers, and dry scrubbers. However, the efficiency of these technologies decreases with a shift from wet to dry scrubbing. Thus, in terms of sulfur dioxide removal efficiency spray towers and packed bed scrubbers tend to have efficiencies in the range of 80 to 90%, semi-dry scrubbers in the range 60 to 80%, and dry scrubbers in the range 50 to 60%.

Cory Environmental has chosen to rely on a semi-dry scrubber system for the removal of acid gases, notwithstanding its slightly lower efficiency when compared to a spray tower or a packed bed scrubber. This is for a number of operational and environmental reasons. Wet scrubbing systems give rise to a wet exhaust gas which requires the deployment of a mist eliminator downstream. Wet systems also give rise to liquid effluent which will require treatment. The EC Directive on the incineration of hazardous waste[8] proposes that the generation of liquid effluents from the operation of an incinerator should be minimized as far as possible. There is a clear steer that systems that give rise to liquid effluents will soon be no longer regarded as BAT. It is probable that this principle will extend to other types of incineration and waste burning in the future.

The flue gas will leave the boiler at approximately 230 °C and will enter the spray absorber through a top-mounted hot air inlet box. The latter will be equipped with profiled baffle plates and adjustable vane rings which will create the conditions necessary for the intimate mixing of the flue gas with the hydrate-of-lime suspension that will be used as the scrubber reagent. The HCl and SO_2 concentrations of the clean gas will be used to control the rate of hydrate-of-lime suspension added when the pollutant concentration in the flue gas is below the base load. For normal to maximum loads the hydrate-of-lime suspension will be added at a constant rate.

The following reactions occur when the atomized liquid comes in contact with the hot flue gases in the absorber:

Gaseous pollutant	Sorbent	Reaction products
2HCl	+ $Ca(OH)_2$	→ $CaCl_2 + 2H_2O$
2HF	+ $Ca(OH)_2$	→ $CaF_2 + 2H_2O$
SO_3	+ $Ca(OH)_2$	→ $CaSO_4 + H_2O$
SO_2	+ $Ca(OH)_2 + \frac{1}{2}O_2$	→ $CaSO_4 + H_2O$

Most reaction products leave the absorber entrained in the flue gas, and will be removed downstream by the baghouse filter. The rest leave the spray absorber through the conical bottom section into a hopper.

The following emission limits are guaranteed by the process equipment supplier: sulfur dioxide, $70\,mg\,(Nm^3)^{-1}$; hydrogen chloride, $30\,mg\,(Nm^3)^{-1}$; and hydrogen fluoride, $1\,mg\,(Nm^3)^{-1}$.

Particulates

The choice of abatement technology for particulate removal depends on the particle size distribution range and the removal efficiency required. Table 4

---

[8] Proposal for a Council Directive on the Incineration of Hazardous Waste (COM (92) a final–SYN 406 of 19 March 1992).

Table 4 Particulate removal methods

Method	Description	Key points	Residue	Suitability for Belvedere*
Settling chamber	Gravity	For large particle prior to secondary abatement	Dust	P
Dry impingement separator	Particles hit baffles and drop out while airs flow around baffle. Relies on particle inertia	Efficient for removal of large particles, e.g. more than 15 μm	Dust	P
Dry cyclonic separators	Inertia separator. A vortex is set up in a chamber and large particles move to the outside wall and drop out	Are precleaners for particles more than 15 μm. Cannot use wet/sticky flue gas. Fluctuations in flow volume and density affect efficiency. Higher loading leads to greater efficiency. Use for high efficiency large particle removal. Problem with re-entrainment. Are cheap and robust	Dust	P
Liquid spraying	Spray directed along gaseous effluent path. Spray droplet and particle size are important. Diffusion occurs	Low cost removal of large particles	Slurry—needs further treatment	P
ESP†-dry	Particles in gas stream are electrically charged and separated from the gas stream. A variation is a 2-stage SP in which gases pass through a corona discharge prior to entering the collector plate area	Very efficient in collecting small particles down to submicron more than 99% efficiency. Can operate in high temperature and flue gas humidity (affects agglomeration). Rapping efficiency is key to good ESP operation. Electrical resistivity of particles is significant factor in ESP design. Collection is a function of gas velocity. Low operating pressure drop. More expensive than filters. Efficiency depends on residence time. Re-entrainment occurs. Are large in size. High capital costs. Low energy requirement	Dust	F/G

Table 4 continued

Method	Description	Key points	Residue	Suitability for Belvedere*
ESP†-wet	Sprays the wet incoming gas stream. Liquid droplets gain electrical charge which then collect particulates in gas stream	More efficient than dry ESP	Effluent needs treatment	F
Fabric filters/ baghouses	Permeable bags which allow the passage of gas but not particulate matter. Filtration occurs by impact, diffusion, gravitational attraction and electrostatic forces. Build up of cake on filter surface acts as a medium itself. Filter cleaned by shaking, pulse or reverse air injection	Effective down to submicron range. Efficiency depends on rate of pressure drop increase. Needs large particles to bridge filter pores and build up cake layer. Needs low moisture and low temperature. Efficiency is 99%. Are very effective. Key factors are bag strength and fan capacity	Dust	G
High	Similar to baghouse filter. Typically made of glass fibre material pleated to increase surface area	Removes extremely small particles. Rapidly clogged by large particles	Dust	P

*Poor—P; Fair—F; Good—G.
†ESP = Electrostatic precipitator.

provides summary details on the main types of particulate arresters. In selecting the technology to be used the type of residues generated is a particular concern. Wet systems, whilst concentrating the particulate in the gas stream, produce an effluent which needs further treatment as well as a sludge which is likely to require dewatering before disposal.

For the Belvedere plant, fabric filter collectors, also known as baghouse, were selected as the means of removing particulates from the gas stream. Fabric filters are capable of maintaining mass collection efficiencies of greater than 99% generally, and greater than 95% for particle sizes of 0.2 μm or less in most applications. These efficiencies are largely insensitive to the physical characteristics of the gas and dust, but depend on the fabric cleaning method, the inlet dust loading, the temperature of the flue gases, and the condensation point of metals.

The three principal methods used to accomplish fabric cleaning are mechanical shaking, reverse air flow, and pulse-jet cleaning. Mechanical shaking is performed off-line, whilst the other two methods can be used on-line. For Belvedere, pulse-jet cleaning has been chosen as the most efficient cleaning method. Measurements will be taken before and after the filter. An increasing pressure drop will result in an increased frequency of pulse-jet cleaning. The dust will be purged by suddenly inflating the bags with pulses of compressed air at 7 bar, causing the dust that has collected on the outside to crumble and fall away.

The fabric filter has been designed to deal with an inlet dust burden of some 5800 mg $(Nm^3)^{-1}$ (dry) at a temperature of up to 220 °C. The fabric material, which will be PTFE, Ryton, or glass fibre, will be selected with reference to its resistance to chemicals and moisture. Removal efficiency is very dependent on the temperature of the flue gas as this relates to the condensation point for metals. The condensation point for most metals (such as compounds of lead, cadmium, chromium, and zinc) is above 300 °C. Hence at the 220 °C temperature the filter is designed to operate at, metals will be in a particulate form and consequently easy to collect. Similarly, heavy metal compounds, particularly chlorides, have condensation points below 300 °C and again will be in particulate form for removal.

Fabric filters can remove a wide range of particle sizes, down to sub-micron sizes, and will continue to work in the event of a power failure as they are passive devices. Our process plant suppliers are prepared to guarantee a total particulate emission level of no more than 10 mg $(Nm^3)^{-1}$ for the Belvedere plant. Set out in Table 1 are the guaranteed emission levels for the Belvedere plant contrasted against current national and EC standards as well as the latest German standard. The high efficiency of the system proposed for the Belvedere plant is clearly demonstrated.

Monitoring

Monitoring of both the internal performance of the plant as well as its effects on the external environment is vital if BATNEEC is to be demonstrated. Therefore three sets of monitoring regimes have been incorporated into the design of the Belvedere plant. These look at, respectively, the performance of the plant, emissions to atmosphere, and effects on the surrounding environment.

To ensure that the plant performs to its design criteria and accordingly behaves in the manner predicted in the Environmental Statement, the following will be monitored:

(i) *Combustion Conditions*—the key factors affecting combustion performance; that is temperature, air, rate of refuse feed, and grate speed will all be monitored continuously. All of these will be interlocked via the management control system to ensure optimal combustion conditions are maintained at all times. In the unlikely event of furnace temperatures falling below 850 °C auxiliary burners will cut in.

(ii) *Boiler Performance*—steam production and the efficacy of the ammonia injection system will be monitored continuously.

(iii) *Air Recirculation*—the electrostatic precipitator will have an automatic control system to maintain the operating voltage in the individual fields as close as possible to the flashover voltage. This will maximize the precipitator efficiency. The primary to secondary air ratio will also be monitored. This ratio determines how much recirculated air is required.

(iv) *Spray Absorber*—HCl and SO_2 concentrations in the clean gas will be monitored after the bag filter but before the ID fan. The data will be used to automatically control the amount of hydrate-of-lime suspension added. The temperature of the clean gas will also be monitored and water added as necessary to cool it.

(v) *Baghouse Filter*—pressure drop measurements will be used to monitor bag performance. Particulate monitoring at the stack will also indicate filter failure.

(vi) *Ash Residue*—regular monitoring of the ash residue will be carried out to check that the combustion process has performed as required.

The following continuous (fixed cycle frequency) emission measurements will be made on the flue gas in the clean gas duct at the stack:

(i) particulates—measured directly by using the Tyndall effect (*i.e.* the amount of light reflected by the particulates);

(ii) carbon monoxide;

(iii) hydrogen chloride;

(iv) sulfur dioxide;

(v) nitrogen dioxide; and

(vi) oxygen content.

It is proposed to use Bran and Luebbe monitors for gaseous emissions. These monitors use wet chemical methods of analysis. In addition, temperature and pressure will be measured continuously, thus allowing the measured pollutants to be calculated with respect to standard conditions, as required by the legislation.

Emissions that cannot be measured continuously will be measured at fixed intervals as specified by the regulatory authorities. It is anticipated that heavy metals, volatile organic carbons, dioxins, furans, and hydrogen fluoride will be measured quarterly. For this purpose additional measurement openings have been provided in the ducts between the ID fan and the chimney.

In addition to these on-site measurements, a routine programme of off-site environmental monitoring will also take place. This will be based on the baseline survey commissioned to establish a description of the environmental conditions in the vicinity of the site prior to development commencing. That study assessed the existing environmental quality of the following: soils and grass; air; ecology; noise.

These measures will be repeated on a regular basis at ten sampling locations within a 5 km radius of the site. These locations have been selected after a detailed analysis of wind and dispersion data relating to the proposed plant.

6 Conclusions

Of the some 137 million tonnes of controlled waste that arise annually in the UK, the majority goes directly to landfill. Other treatment and disposal technologies, notably incineration, have played only a modest role in the waste management scene. Indeed, after a burst of enthusiasm in the early 1970s, most incinerators have been allowed to decline such that few are capable of meeting current emission standards without extensive, and expensive, retrofitting.

This picture is changing with developments in legislation forcing greater emphasis on recycling, including maximizing recovery of the energy value of wastes, and setting tougher environmental standards. These developments are all serving to mitigate against landfill as the preferred disposal route, except where geological and environmental conditions are ideal. Rather they are seen instead to be encouraging new investment in alternative waste management technologies, of which waste-to-energy is seen to be of great interest.

The flagship development of Cory Environmental is the proposed waste-to-energy plant at Belvedere in the London Borough of Bexley. Should permission be granted, this plant would deal with 1.2 million tonnes of London's household and commercial waste, thereby doing much to relieve the ever increasing critical shortage of landfill to the south and east of the capital. The plant will export some 103 MW of electricity to the national distribution network, making it the largest waste recycling unit in the UK. The plant enjoys substantial support from Government through its inclusion in the latest tranche of the Non-Fossil Fuel Obligation. The Belvedere plant represents 22.5% of this tranche and 39.4% of the waste band.

The plant has been designed to represent the very best in waste-to-energy technology and to set a standard that few, if any, comparable plants worldwide can achieve. This starts right at the delivery of waste to the plant where up to 1 million tonnes can be delivered in sealed containers on barges. This is an environmentally more acceptable means of transporting waste than conventional road transport.

The technologies employed at the plant to ensure its emissions to the environment are as low as can be achieved practically also represent the latest developments worldwide. The key technologies employed include, firstly, flue gas recirculation. By recirculating some 17% of the exhaust gases back as secondary air into the furnace the thermal efficiency of the process is increased and the creation of pollutants, particularly NO_x, is minimized. To further reduce NO_x emissions ammonia is also injected into the boiler. The net effect of this treatment system will be a reduction of NO_x by approximately 20% and an improvement in boiler efficiency of between 2 to 3%.

Secondly, measures have been taken specifically both during and after combustion to prevent the formation, reformation, and release of PCDDs and PCDFs. Combustion conditions have been designed so that maximum burnout of waste is achieved with consequently only a minimum amount of products of incomplete combustion, the precursors of PCDDs/PCDFs, being formed. In addition, residual PCDDs/PCDFs will be removed from the gas stream by a combination of gas scrubbing and activated carbon injection technologies. These measures, collectively, allow the technology suppliers to guarantee an emission level for PCDD/PCDF of $0.1 \, \text{ng} \, (\text{Nm}^3)^{-1}$ as against the national standard of $1 \, \text{ng} \, (\text{Nm}^3)^{-1}$. The injection of activated carbon into the gas stream has also been proven to be effective in removing cadmium and mercury. In addition, the combination of proven technologies to be deployed at Belvedere allow the process plant suppliers to guarantee emission levels considerably better than current national and international standards.

An extensive programme of monitoring is proposed. The performance of the plant will, as far as practicable, be continuously monitored. The results of this monitoring will be fed electronically into the management control system governing the operation of the plant. In that way the maintenance of the plant's performance at the optimum can be guaranteed. To supplement this in-plant monitoring an extensive programme of off-site monitoring is also proposed. In that way the effects of the plant on the surrounding environment will be assessed regularly and the validity of the environmental assessment originally made for the proposed plant checked.

The Belvedere waste-to-energy plant therefore represents one of the first of a new generation of technologies to deal with wastes in an environmentally sensitive and benign way. A high degree of control and predictability lies at the heart of this technology, making its environmental effects minimal and obvious.

Organic Micropollutant Emissions from Waste Incineration

G. H. EDULJEE

1 Introduction

Background

Concern over organic micropollutant emissions from waste incinerators dates from the 1960s, following the identification of a range of polyaromatic hydrocarbons (PAHs) and other organic species in stack emissions from municipal solid waste (MSW) incinerators.[1] The US Environmental Protection Agency (US EPA) were in the forefront of regulating incinerator operating conditions specifically for the purpose of minimizing organic emissions, sponsoring laboratory research that led to the formulation of minimum temperature and gas phase residence time requirements of 1200 °C $\pm$ 100 °C and 2 s respectively, when incinerating polychlorinated biphenyl (PCB) wastes.[2] These conditions were subsequently adopted by other national regulatory bodies to control the incineration of hazardous wastes. However, public anxiety over the potential health risk posed by emissions of organic micropollutants underwent a step change in the late 1970s, following the identification of polychlorinated dibenzo-p-dioxins (PCDDs) and dibenzofurans (PCDFs) in incinerator emissions, coinciding with the release of 2,3,7,8-TCDD and subsequent environmental contamination at Seveso. From the mid-1980s, additional and increasingly stringent controls were placed on incineration operations; for example by setting emission limits for PCDDs and PCDFs, total organic carbon (TOC), and carbon monoxide (CO).

The notoriety of PCDDs and PCDFs has displaced almost every other potential emittant from the public stage, causing incineration to be reviled with a passion not shown for other waste disposal options, and spawning a veritable explosion of research effort on the fate and environmental effects of these compounds following release into the environment.[3] Among currently operating plant, these compounds, along with PCBs, have been the subject of great public concern, leading to the commissioning of official studies to investigate allegations

[1] R. P. Hangebrauck, D. J. von Lehmden, and J. E. Meeker, 'Sources of Polynuclear Hydrocarbons in the Atmosphere', US EPA Report 999-AP-33, 1967.
[2] US EPA, 'Polychlorinated Biphenyls (PCBs) Manufacturing, Processing, Distribution in Commerce, and Use Prohibitions', Federal Register, 44, 31514, May 31, 1979.
[3] G. H. Eduljee, Chem. Br., 1988, 24, 1223.

71

of ill-health and environmental damage.[4-6] None of these studies have established incineration as the cause of the alleged damage, but nevertheless it remains an unpopular and controversial waste management option.

For the most part, this chapter treats organic micropollutants as a generic class of compounds, rather than concentrating on the few chemicals that have been of concern: the scientific literature abounds with texts and papers on PCBs, PCDDs, and PCDFs and their effect on humans and the environment. A discussion of the thermochemical processes that control the formation of organic micropollutants is followed by an examination of some of the design and operational techniques that can be applied to minimize and control these emissions. The formation of PCDDs and PCDFs is presented as an example of the reactions that can occur in the cooler, post-combustion zone of an incinerator. Finally, the health implications of exposure to organic micropollutants are discussed, again treating organic micropollutants in their totality. Throughout, an attempt is made to link the chemistry and mechanisms of formation to the design and operation of full-scale plant.

Products of Combustion

While carbon dioxide and water are the principal products of combustion of wastes, other compounds can be formed depending on the composition of the waste. For example, chlorinated wastes will generate hydrocarbon chloride, sulfur-bearing wastes will produce sulfur oxides, and the metallic constituents will be transformed to their respective oxides, attaching either to the ash or combining with light non-combustible material to be emitted into the atmosphere via the stack, as particulate matter. All combustion sources also emit small amounts of organic matter, comprising the uncombusted fraction of organic compounds present in the waste feed, and the reaction products from interactions between various organic species (see Section 2). Table 1[7,8] presents typical compositions of raw and scrubbed flue gases from MSW incinerators, indicating the relative proportions of the main emittants. The majority of the metals emissions are associated with the particulate phase.

Analysis of bottom-ash from the grate, flyash from the gas cleaning plant, and flue gas from MSW incinerators has indicated that about 1% of the carbon entering an incinerator leaves the combustor in the bottom-ash, 0.1% is associated with the flyash and some 0.01% is emitted in the flue gas in the form of

4 Welsh Office, 'The Incidence of Congenital Malformations in Wales, with particular reference to the District of Torfaen, Gwent', Welsh Office, Cardiff, 1985.
5 Scottish Office, 'Bonnybridge/Denny Morbidity Review', Scottish Office, Edinburgh, 1985: 'Report of a Working Party on Microphthalmos in the Forth Valley Health Board Area', Scottish Office, Edinburgh, 1988.
6 Welsh Office, 'Panteg Monitoring Report', Second Report to the Welsh Office by the Environmental Risk Assessment Unit, University of East Anglia, Welsh Office, Cardiff, 1993.
7 P. Clayton, P. Coleman, A. Leonard, A. Loader, I. Marlowe, D. Mitchell, S. Richardson, D. Scott, and M. Woodfield, 'Review of Municipal Solid Waste Incineration in the UK', Report LR 776 (PA), Warren Spring Laboratory, Stevenage, 1991.
8 L. Barniske, *Waste Manage. Res.*, 1987, **5**, 347.

Table 1 Typical composition of raw and scrubbed flue gases, NTP[7,8]

	Raw gas	Scrubbed gas/mg m^{-3}
Water	10–18% by volume	—
Carbon dioxide	6–12% by volume	—
Oxygen	7–14% by volume	—
Particulate matter	2–10 g m^{-3}	20–30
Hydrochloric acid	250–2000 mg m^{-3}	10–30
Hydrofluoric acid	0.5–9 mg m^{-3}	0.5–2
Sulfur oxides	200–1000 mg m^{-3}	200–300
Nitrogen oxides	100–400 mg m^{-3}	100–400
Carbon monoxides	50–100 mg m^{-3}	50–100
Total organic carbon	<20 mg m^{-3}	<20

organic micropollutants,[9] the remaining carbon being converted to carbon dioxide. Concentrations of TOC average 10 g kg^{-1} in the bottom-ash, 40 g kg^{-1} in flyash, and 20 mg Nm^{-3} in the flue gas. This chapter is concerned with emissions of organic micropollutants to atmosphere, though the presence of organics in bottom- and flyash has in recent years been raised as an issue in relation to the disposal of these products in landfills.

2 Sources, Composition, and Levels

Sources

There are essentially three routes by which organic micropollutants can be formed and emitted from incineration processes:

  (i) *As a result of incomplete combustion of organic wastes present in the original waste.* For example, if PCB is subjected to a destruction and removal efficiency (DRE) of 99.9999%, then the uncombusted fraction comprising 0.0001% of the original feedstock (or 1 mg for every kilogramme incinerated) will be emitted to atmosphere via the stack.
 (ii) *As a result of the synthesis of 'new' compounds in the combustion and post-combustion zones of an incineration plant.* An example is the formation of polychlorinated dibenzo-*p*-dioxins (PCDDs) and polychlorinated dibenzofurans (PCDFs) in the boiler and gas cleaning components of an incineration plant.
(iii) *Organic compounds brought into the incineration system from other sources* such as combustion air, scrubber, water, and support fuel.

The US regulatory system differentiates between emissions arising from the three routes described above. The Resource Conservation and Recovery Act of 1976 (RCRA) contains, in Appendix VIII of the Act, a list of over 400 inorganic and organic hazardous chemicals, from which representative chemicals known as Principal Organic Hazardous Constituents (POHCs) are chosen to best represent the particular waste stream to be incinerated. In order to receive a

[9] P. H. Brunner, M. D. Muller, S. R. McDow, and H. Moench, *Waste Manage. Res.*, 1987, **5**, 355.

G. H. Eduljee

RCRA permit, the operator of an incineration facility is required to demonstrate a DRE of 99.99% for each POHC in the waste feed. Emissions arising from route (i) are therefore classed as POHC emissions. Organic emissions arising from routes (ii) and (iii) are termed Products of Incomplete Combustion (PICs) but the definition is restricted to Appendix VIII compounds present in the stack gas, but not found above $100\,\mu g\,g^{-1}$ in the waste.[10] PIC emissions are not subject to regulation in the US, although the US EPA have proposed that they be limited to 0.01% of the POHC input to hazardous waste incinerators.[11] It has been noted[12] that, as presently defined, organic chemicals which are detected in the stack gas but are not listed in Appendix VIII can neither be classed as POHCs nor as PICs; the result is that a large percentage of stack emissions remain unclassified, possibly larger than the combined Appendix VIII constituents that have been identified. The US terminology has increasingly been adopted by regulators and the waste management industry in Europe, but generally without an appreciation of the specific regulatory context. The tendency in Europe has been to define *all* organic compounds present in the stack gas as PICs, irrespective of their origin or mechanism of formation, an interpretation that is more appropriate from the standpoint of the risk to public health. This chapter retains the US definitions insofar as they differentiate between emissions via route (i) (POHC emissions) and route (ii) (PIC emissions) above.

Composition

Studies on the composition of the organic fraction of stack gas emissions from operating incinerators have generally concentrated on PAH, PCB, PCDD, and PCDF emissions, with municipal solid waste (MSW) and hazardous waste incinerators receiving the most attention. The most comprehensive databases for a wider range of organic emissions are US EPA-sponsored laboratory and full-scale combustion trials on sewage sludge incinerators[13-15] and hazardous waste combustors,[10] and a Canadian study on emissions from pilot-scale and full-scale MSW incinerators.[16] The US studies have generally been restricted to Appendix VIII POHCs and PICs. The range and concentrations of POHC emissions will vary from facility to facility, depending on the waste being incinerated. Thus, for a plant incinerating pharmaceutical wastes, the POHCs are likely to include the unburnt fractions of chemicals which are specific to this

[10] A. Trenholm, P. Gorman, and G. Jungclaus, 'Performance Evaluation of Full-Scale Hazardous Waste Incinerators', Vols I-V, US EPA, EPA-600/2-84-181a–e, Cincinnati, 1984.
[11] US EPA, 'Incinerator Standards for Owners and Operators of Hazardous Waste Management Facilities; Interim Final Rule and Proposed Rule', Federal Register, 46, 7666, January 23, 1981.
[12] C. C. Travis and S. C. Cook, 'Hazardous Waste Incineration and Human Health', CRC Press, Florida, 1989, p. 101.
[13] D. A. Tirey, R. C. Striebich, B. Dellinger, and H. E. Bostian, *Haz. Waste Haz. Mater.*, 1991, **8**, 201.
[14] Radian Corporation, 'Emission Test Report Sewage Sludge Test Program, Sites 1–4', for US EPA Contract No. 68-02-6999, 1987.
[15] Radian Corporation, 'Final Test Reports—Sites 01–03, Sewage Sludge Incinerator SSI-A National Dioxin Study, Tier 4, Combustion Sources', for US EPA Contract No. 68-03-3148, 1986.
[16] A. Finkelstein, R. Klicius, and D. Hay, in 'Emissions from Combustion Processes: Origin, Measurement, Control', ed. R. Clement and R. Kagel, Lewis Publishers, Boca Raton, 1990, p. 243.

74

waste stream and the particular site. Since most regulatory systems require a minimum DRE ranging from 99.99% to 99.9999% for each POHC in the waste feed, nominal concentrations of the relevant compounds in the stack gas can be calculated if the feed has been adequately characterized: this can be generalized to encompass any organic chemical present in the waste feed, irrespective of its presence on the Appendix VIII list. In practice, it has been found that the DRE decreases with falling concentrations in the waste feed.[17]

Of greater interest are emissions of 'new' chemicals formed during the combustion process. The most detailed studies of PIC emissions, conducted in the US, are constrained by the regulatory framework discussed above. It has been estimated[10] that the 35 Appendix VIII PICs detected in the stack gases of eight hazardous waste incinerators represented approximately 1% of the total unburnt hydrocarbons, the unidentified fraction being non-chlorinated C_1–C_5 hydrocarbons, principally methane and ethylene.[18] An alternative estimate is that test programmes have been able to identify approximately 60% of the organic emissions from hazardous waste incinerators.[19] It is therefore difficult to provide accurate estimates of emissions of PICs; using US terminology, the total PIC emission rate (for both Appendix VIII and non-Appendix VIII compounds) can range from one-tenth to ten times the emission rate for POHCs.

A striking feature of the studies on PIC emissions is that many of the same organic compounds have been detected in almost all stack gas samples, irrespective of the waste feed or type of combustor. This suggests a dominant and common mechanism of formation; namely, the recombination of molecular fragments which result from pyrolysis or partial oxidation of the constituents in the waste feed. A discussion of this mechanism is presented in Section 3 below. Table 2 provides a list of the PICs identified in emissions from incinerators, boilers, and industrial furnaces burning hazardous wastes, ranked in order of decreasing concentration in the stack gas.[18]

Also included in the table are data on emissions from sewage sludge incinerators,[20] illustrating the similarity in both the composition and relative amounts of the PICs detected in stack gases, above a detection limit of $100 \, \text{ng m}^{-3}$. Data for C_1 and C_2 hydrocarbons, while based on emissions from fossil fuel combustion, are compatible with trials on a hazardous waste incinerator in which methane and ethylene accounted for 33–97% of the identified organic emissions.[21]

Extensive sampling of emissions from a large pilot-scale MSW incinerator was conducted by Environment Canada under the National Incinerator Testing Program (NITEP).[16] Organic emissions testing was confined to PCDDs, PCDFs, PAHs, PCBs, chlorobenzenes, and chlorophenols. A similar suite of

---

[17] C.C. Travis and S.C. Cook, 'Hazardous Waste Incineration and Human Health', CRC Press, Boca Raton, 1989, p. 40.

[18] C.R. Dempsey and E.T. Oppelt, *Air Waste*, 1993, **43**, 25.

[19] US EPA, 'Guidance on PIC Controls for Hazardous Waste Incinerators, Draft Final Report', Contract No. 68-01-7287, Washington, 1989.

[20] US EPA, 'Technical Support Document—Incineration of Sewage Sludge', Office of Water Regulations and Standards, Washington, 1989.

[21] US EPA, 'Total Mass Emissions from a Hazardous Waste Incinerator', Final Report, EPA-600/S2-87/064, Cincinnati, 1987.

Substance	Hazardous waste incinerators[a,b] (percentage of total)	Sewage sludge incinerators (percentage of total)
C_2 Hydrocarbons	44.804	61.414
C_1 Hydrocarbons	25.301	31.560
Acrylonitrile	Not detected	1.840
Benzene	12.988	0.556
2,4-Dinitrophenol	Not detected	0.469
Methylene chloride	4.626	0.150
Chloroform	3.708	0.380
Chloromethane	2.127	0.0018
1,2-Dichloroethane	1.881	0.380
Toluene	1.452	0.386
Tetrachloroethylene	0.783	0.411
2,4,5-Trichlorophenol	0.380	Not detected
Carbon tetrachloride	0.262	0.0033
o-Dichlorobenzene	0.250	Not tested
p-Dichlorobenzene	0.227	Not tested
Trichloroethylene	0.216	0.216
bis(2-Ethylhexyl) phthalate	0.205	0.085
1,2,4-Trichlorobenzene	0.203	0.054
1,1,1-Trichloroethane	Not tested	0.175
1,1,2-Trichloroethane	0.097	0.034
Methyl ethyl ketone	0.087	0.312
Phenol	0.087	0.214
1,1-Dichloroethylene	0.083	0.0018
Diethyl phthalate	0.082	0.064
1,1,2,2-Tetrachloroethane	0.045	Not detected
Vinyl chloride	0.037	Not tested
Pentachlorophenol	0.025	0.139
Hexachlorobenzene	0.024	Not tested
1,1-Dichloroethane	0.0089	Not tested
Bromomethane	0.0056	0.0022
Dichlordifluoromethane	0.0032	Not detected
Benzo(a)anthracene	0.0029	Not tested
2,4-Dichlorophenol	0.0013	0.072
Acetonitrile	0.00069	0.532
2,3,7,8-Hexachlorodibenzo-p-dioxins	0.0000080	0.00000092
Other hexachlorodibenzo-p-dioxins	0.0000187	0.0000040
2,3,7,8-Heptachlorodibenzo-p-dioxins	0.0000055	0.00000093
Other heptachlorodibenzo-p-dioxins	0.0000055	0.0000011
2,3,7,8-Tetrachlorodibenzo-p-dioxin	0.0000041	0.00000014
Other tetrachlorodibenzo-p-dioxins	0.00016	0.000013
2,3,7,8-Tetrachlorodibenzofuran	0.0000037	0.0000066
2,3,7,8-Pentachlorodibenzo-p-dioxins	0.0000033	0.00000025
Other pentachlorodibenzo-p-dioxins	0.000067	0.0000061
Ethylbenzene	Not tested	0.087
Benzo(a)pyrene	Not detected	0.031
Hexachloroethane	Not detected	0.0014
PCBs	Not detected	0.00036

Table 2 Emissions of organic micropollutants from hazardous waste and sewage sludge incinerators[18,20]

[a]Total organic micropollutant emissions: 37 943 $\mu g\,m^{-3}$.
[b]Incineration of non-PCB wastes.
[c]Total organic micropollutant emissions: 55 687 $\mu g\,m^{-3}$.

Table 3 Summary of organic micropollutant emissions from Canadian and Norwegian tests[16,23]

	Canadian tests $\mu g\,Nm^{-3}$	Norwegian tests $\mu g\,Nm^{-3}$
Chlorobenzenes	3.3–9.9	0.034–3.8
PAHs	3.2–21.9	0.84–6000
PCBs	1.7–7.0	<0.00003–0.06
PCDDs/PCDFs	0.063–0.597	0.047–1.8
Chlorophenols	5.1–23.7	ND

ND = not determined

compounds was examined in the stack gases of Swedish MSW incinerators[22] and small MSW incinerators in Norway.[23] A summary of the Canadian and Norwegian emissions data is provided in Table 3.

While the US, Canadian, and Norwegian data provide an indication of the range and relative proportions of organic micropollutants emitted from combustors, it should be noted that the emission concentrations measured during the trial burns relate to plant and equipment that pertained prior to the introduction of more advanced gas cleaning equipment. In response to increasingly stringent emission limits imposed on incinerators from the mid-1980s, air pollution control techniques such as the injection of activated carbon into the gas stream have been introduced, primarily to reduce emissions of trace metals, and organics such as PCDDs and PCDFs. Since these techniques also remove other organic species from the gas stream, pre-1990 emissions tests are unlikely to be representative of the emission concentrations of organic micropollutants that can routinely be achieved on modern incineration plant. For example, emissions of chlorinated benzenes, chlorinated phenols, and PAHs from Swedish MSW incinerators were anticipated to fall by a factor of 10–100 from a 1985 base, following the implementation of a programme of optimization coupled with more stringent emission limits.[22] All of the data presented in Tables 2 and 3 relate to trial burns conducted between 1983 and 1987.

The US EPA trial burns also identified organic micropollutants in the stack gas, that were introduced into the combustion system via the scrubber water, auxiliary fuel, or ambient air.[24] For example, in one test chloroform was detected in the scrubber make-up water at a concentration of $100\,\mu g\,l^{-1}$. The scrubber effluent contained less than $1\,\mu g\,l^{-1}$ of chloroform, the stripped chemical entering the stack gas and contributing $70\,\mu g\,m^{-3}$ (about 14%) of the total mass of PICs detected, excluding C_1 and C_2 hydrocarbons. Ambient air, introduced as combustion air or via leaks in the ductwork, accounted for some of the halogenated organics detected in the stack gas, though this source was not a significant contributor. Finally, auxiliary fuel used at the test facility also contained six volatile halogenated hydrocarbons at concentrations greater than (but approaching) $100\,\mu g\,g^{-1}$, compounds which were also detected in the stack

[22] U. G. Ahlborg and K. Victorin, *Waste Manage. Res.*, 1987, **5**, 203.
[23] Ch. Benestad and M. Oehme, *Waste Manage. Res.*, 1987, **5**, 407.
[24] A. Trenholm, R. Hathaway, and D. Oberacker, in 'Incinerating Hazardous Wastes', ed. H. M. Freeman, Technomic Publishing Co. Inc., Lancaster, Pennsylvania, 1988, p. 35.

gas. The significance of auxiliary fuel as a source of organic micropollutants independent of the waste stream, is that low concentrations of chemicals in the feed material have been associated with lower DREs, and hence the contribution to organic emissions of a chemical present in auxiliary fuel may be disproportionately high relative to its concentration in the fuel.

However, the contribution of this latter source of organic micropollutants should be kept in perspective. Emissions from waste-fired and fossil fuel-fired combustors have been shown to be qualitatively similar in composition. With C_1 and C_2 hydrocarbons and seven other chemicals dominating the total organic emissions, changes in the relative proportions of the remaining organic micropollutants will affect a small percentage (in the order of 4%) of the organic fraction.

3 Mechanisms of Formation

Introduction

The objective of incineration is to attain complete combustion of the waste, converting carbon, hydrogen, and other constituents such as halogens to carbon dioxide, water, and hydrogen halides. The following general stoichiometry can be assigned to the reaction:

$$C_xH_yCl_z + \left(x - \frac{y-3}{4}\right)O_2 = xCO_2 + zHCl + \frac{y-3}{z}H_2O \tag{1}$$

The required conditions are oxidative; sufficient oxygen should be supplied to the combustor to convert the carbon to carbon dioxide, and to combine with the hydrogen remaining after HCl formation, to generate water. If the waste contains insufficient hydrogen to effect the conversion of halogens to hydrogen halides, then additional hydrogen is supplied through the introduction of auxiliary fuel. In order to maintain oxidative conditions, incinerators invariably operate with an excess of oxygen, typically 50% to 100% in excess of stoichiometric requirements.

Reaction between the organic constituents of the waste and oxygen in the combustion air occurs in the flame zone, and in the post-flame environment. Gas phase residence times in the flame zone are in the order of microseconds, during which time over 95% of the organic chemicals are oxidized.[25] Residence times in the post-flame zone vary between 0.5 and 2 s, depending on the type of waste and the corresponding regulatory requirements. Under oxidative conditions, exothermic decomposition reactions drive the process to completion, with the formation of the products depicted in Equation (1). However, incinerators are not perfect combustors: channelling and layering of gases, perturbations in temperature, and variations in the reactivity of the waste constituents result in localized pockets of gaseous reactants in which oxygen is depleted to below the stoichiometric minimum. In these pockets, decomposition of the organic constituents proceeds through pyrolytic processes. While oxidative kinetics dominate the overall global decomposition of an organic constituent, emissions

[25] W.R. Seeker, W.S. Lanier, and M.P. Heap, 'Municipal Waste Combustion Study: Combustion Control of Organic Emissions', US EPA Report No. EPA/530-SW-87-021c, Cincinnati, 1987.

of organic micropollutants correlate with pyrolytic reaction pathways and associated decomposition products.[13,26,27]

This section discusses the global reaction kinetics of the destruction of organics during incineration, the mechanisms that dominate the pyrolytic decomposition pathways, and, in view of their toxicological significance, the mechanism of PCDD and PCDF formation in incinerators.

Global Decomposition Kinetics

The global reaction mechanism for thermal decomposition must take into account two reaction pathways, those for oxidation and pyrolysis:[28]

$$-\frac{d[X]}{dt} = k_1[X]^a + k_2[X]^b[O_2]^c \qquad (2)$$

k_1 and k_2 are the global rate constants for pyrolysis and oxidation, $[X]$ and $[O_2]$ the concentrations of the organic species and of oxygen, and a, b, and c are the reaction orders for the decomposition of species X. Discounting the pyrolytic pathway owing to the presence of excess oxygen in the system and expressing the temperature dependence of the rate constants by the Arrhenius equation, the following expression is obtained, linking the fraction of the parent species remaining (f), with the temperature of combustion (T):

$$T = 120E\left[\ln\left(\frac{-tA}{\ln f}\right)\left(\frac{O_2}{0.21}\right)^c\right]^{-1} \qquad (3)$$

f is the fraction of species X remaining after combustion at temperature T (K); O_2 is the fraction of oxygen in the reaction atmosphere; E is the activation energy (kJ mol^{-1}); A is the Arrhenius coefficient (s^{-1}), and t is the residence time (s) of species X at the combustion temperature. Regression analysis of thermal decomposition profiles determined in laboratory tests indicated that first order reaction kinetics best represented the data. Calculated combustion temperatures required for 99.99% destruction of a chemical held in the combustor for a residence time of 2 s, ranged from 600 °C for 1,1,1-trichloroethane, to 650 °C for hexachloroethane, and 910 °C for acetonitrile and acrylonitrile. The effect of perturbations in combustion conditions was also examined.[28] Even if only 1% of the organic constituent experiences, say, a residence time of 0.2 s and 1% oxygen (as opposed to the global test conditions of 2 s and 8% oxygen), the destruction efficiency falls from 99.99% to 96%, resulting in emissions of the uncombusted fraction that are two orders of magnitude higher than if the test conditions had been uniformly applied.

Mechanisms of PIC Formation

Homogeneous Gas Phase Reactions. Accepting the premise that localized pyrolytic conditions are the prime cause of PIC formation, reaction pathways

[26] C. C. Lee, G. L. Huffman, and S. M. Sasseville, *Haz. Waste Haz. Mater.*, 1990, **7**, 385.

[27] J. L. Graham, D. L. Hall, and B. Dellinger, *Environ. Sci. Technol.*, 1986, **20**, 703.

[28] B. Dellinger, J. L. Torres, W. A. Rubey, D. L. Hall, and J. L. Graham, *Haz. Waste*, 1984, **1**, 137.

Table 4 Some examples of thermal decomposition reactions[26]		*Comments*
Unimolecular reactions		
(i) Bond homolysis $CCl_4 \rightarrow CCl_3 + Cl$		Breaking of the weakest bond
(ii) Three-centre elimination $CHCl_3 \rightarrow CCl_2 + HCl$		Elimination of stable species such as HCl
(iii) Four-centre elimination $C_2H_4Cl \rightarrow C_2H_4 + HCl$		Typical decomposition pathway of organics with carbons connected by a single bond
(iv) Six-centre elimination $C_6H_4(COOC_2H_5)_2 \rightarrow C_6H_4$ $(COOH)_2 + 2C_2H_4$		Typical of phthalates, sulfates, and esters
Bimolecular reactions		
(i) Atom metathesis $CH_3Cl + H \rightarrow CH_2Cl + H_2$		Typical of halogenated alkyls
(ii) Electrophilic addition $C_2H_4 + OH \rightarrow C_2H_4OH$		Other typical radicals include H and Cl
(iii) Hydrogen abstraction $CH_3Cl + OH \rightarrow CH_2Cl + H_2O$		Other typical radicals include Cl and H
(iv) Displacement $C_6H_4Cl_2 + H \rightarrow C_6H_5Cl + Cl$		Displacement of H by Cl is not favoured

can be proposed that lead to formation of intermediate organic species and stable PICs. Homogeneous gas phase reaction mechanisms can be divided into two broad groups, those involving unimolecular decomposition, and those involving bimolecular processes.[26,29] Table 4 summarizes some typical reactions within these two groups. While unimolecular reactions have a role in the thermal decomposition of organics, bimolecular reaction in the form of radical attack is believed to be the predominant mechanistic pathway for both complete oxidation and PIC formation.

The decomposition pathways followed by organic constituents of the waste

[29] W. Tsang, in 'Incineration of Hazardous Waste—Toxic Combustion By-Products', ed. W. R. Seeker and C. P. Koshland, Gordon and Breach Science Publishers, Philadelphia, 1989, p. 99.

involve a complex array of reactions. A relatively simple example of the decomposition of methyl chloride (CH_3Cl) is discussed below to illustrate PIC formation through some of the reactions listed in Table 4.[30] The decomposition is initiated by two dominant reactions: bond homolysis through cleavage of the weak C—Cl bond to form the Cl radical, and the formation of the hydroxyl radical (OH) by the reaction $H + O_2 \rightarrow OH + O$. Attack by these radicals on CH_3Cl results in hydrogen abstraction and the formation of the CH_2Cl radical. Decomposition of this radical proceeds through a series of recombination reactions:

$$CH_2Cl + CH_2Cl \rightarrow C_2H_3Cl + HCl \tag{4}$$
$$CH_2Cl + CH_2Cl \rightarrow CH_2ClCH_2 + Cl \tag{5}$$
$$CH_2Cl + CH_2Cl \rightarrow CH_2ClCH_2Cl \tag{6}$$
$$CH_2ClCH_2Cl \rightarrow C_2H_3Cl + HCl \tag{7}$$

The stable products of decomposition are HCl, C_2H_3Cl, and $1,2\text{-}C_2H_4Cl_2$. Further radical attack results in the formation of ethene (CH_2CH_2):

$$CH_2ClCH_2 \rightarrow CH_2CH_2 + Cl$$
$$CH_3Cl + H \rightarrow CH_3 + HCl$$
$$CH_3 + CH_2Cl \rightarrow CH_2CH_2 + HCl$$

Other reactions leading to the formation of ethyne (CHCH), CO, and CO_2 have also been identified; for example:

$$C_2H_3Cl \rightarrow CHCH + HCl$$
$$CH_2CH_2 + (H,Cl) \rightarrow CH_2CH + (H_2, HCl)$$
$$CH_2CH \rightarrow CHCH + H$$

Benzene can be formed through reactions involving C_4H_2 and C_4H_4 precursors:

$$CH_2CH + C_2HCl \rightarrow C_4H_4 + Cl$$
$$C_4H_4 + (H,Cl) \rightarrow C_4H_3 + (H_2, HCl)$$
$$C_4H_3 \rightarrow C_4H_2 + H$$

Addition of C_2 species to the C_4 species, followed by cyclization and dehydrogenation of the intermediates, results in the formation of benzene (C_6H_6).[31] Chlorination of benzene could then lead to the formation of chlorobenzenes as PICs.

An interesting observation has been made in respect of the role of soot in the post-flame environment of a combustor.[32] A proportion of the PICs formed in the combustor were found to be adsorbed onto, or trapped interstitially in the soot generated in an oxygen-deficient environment. While the PICs present in the gas phase exhibited oxidative decomposition kinetics as represented in Equations (2) and (3) above, the PICs absorbed on or trapped in the soot were governed by entirely different destruction kinetics associated with the combustion on coke and graphite surfaces.

The above reaction mechanisms are of more than theoretical interest. Oxygen-deficient, fuel (*i.e.* hydrogen)-lean conditions will favour recombination reactions

[30] E. M. Fisher and C. P. Koshland, *Combust. Flame*, 1992, **90**, 185.
[31] M. Qun and S. M. Senkan, *Haz. Waste Haz. Mater.*, 1990, 7, 55.
[32] R. D. VanDell and N. H. Mahle, in 'Emissions from Combustion Processes: Origin, Measurement, Control', ed. R. Clement and R. Kagel, Lewis Publishers, Boca Raton, 1990, p. 93.

(4) and (6), leading to the formation of C_2H_3Cl as a PIC, together with other PICs represented by subsequent reaction sequences. In fuel-rich environments, H and CH_3 radicals will dominate, leading to the formation of ethene rather than C_2H_3Cl.[30] Soot can also be formed under both these conditions. On the other hand, an abundance of oxygen will result in decomposition of ethene and other carbonaceous intermediates to CO and ultimately to CO_2, the desired end product, and all chlorine is converted to HCl, which is scrubbed out of the gases. Further, the reactions are also temperature dependent, with overall destruction of the parent compound decreasing with a fall in temperature, and PIC formation under oxygen-deficient conditions initially increasing with a rise in temperature, and then falling as thermal decomposition dominates.[27] Thus, careful design of the combustor is essential to ensure good mixing and to eliminate localized cold spots and micro-environments in which pyrolytic or oxygen-deficient decomposition pathways are favoured.

Heterogeneous Reactions. Following the identification of PCDDs and PCDFs in incinerator emissions,[33] considerable research effort has been put into the elucidation of the key reactions that govern their formation. Three pathways have been proposed:

(i) PCDD and PCDF emissions occur as a result of their presence in the waste feed (*i.e.* as the fraction of uncombusted chemical).
(ii) PCDD/Fs are produced by reactions between organic precursors such as chlorobenzenes, chlorophenols, PCBs, *etc.*
(iii) PCDD/Fs are formed via *de novo* synthesis, from organic fragments generated during thermal decomposition of organics, and organic or inorganic chlorine donors.

Pathway (i) has been addressed above; mass balance experiments have discounted this as a significant source of PCDD/Fs, since their emissions often exceed the maximum release rate calculated on the basis of the feed concentration, pointing to a genuine PIC formation mechanism. Homogeneous gas phase reaction mechanisms have been shown to generate PCDD/Fs from almost any starting material containing carbon, hydrogen, and chlorine,[29,34] but whereas at high temperatures typical of the combustion zone precursor decomposition dominated over potential PCDD/F formation reactions, at lower post-combustion temperatures the rate of formation could not account for the observed concentrations.

Subsequent work has demonstrated that the dominant mechanism of formation involves heterogeneous, surface-catalysed reactions between chlorinated precursors and/or the products of *de novo* synthesis, on flyash particles held in the relatively cool (200 °C–400 °C) post-combustion environment of the boiler or particulate arrestment equipment.[35,36] A general scheme for the formations of PCDDs and PCDFs can be depicted as in Figure 1.

[33] K. Olie, P. L. Vermeulen, and O. Hutzinger, *Chemosphere*, 1977, **6**, 454.
[34] G. Eklund, J. R. Pedersen, and B. Stromberg, *Chemosphere*, 1988, **17**, 575.
[35] H. Vogg and L. Stieglitz, *Chemosphere*, 1986, **15**, 1373.
[36] H. Hagenmaier, M. Kraft, H. Brunner, and R. Haag, *Environ. Sci. Technol.*, 1987, **21**, 1080.

Figure 1 Schematic representation of the post-combustion formation of PCDDs and PCDFs

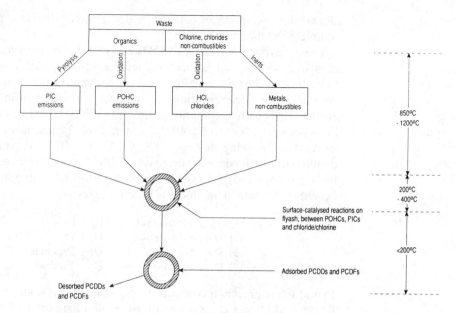

The key requirements for PCDD/F formation are an oxygen-rich environment, a source of chlorine, and the presence in the flyash, of a metal capable of catalysing a Deacon reaction. *De novo* synthesis is initiated by the formation of chlorine via the action of oxygen on hydrogen chloride:

$$MCl_2 + 1/2O_2 \rightarrow MO + Cl_2$$
$$MO + 2HCl \rightarrow MCl_2 + H_2O$$

leading to

$$2HCl + 1/2O_2 \rightarrow H_2O + Cl_2$$

Hydrogen chloride is formed as a decomposition product of chlorinated organics, and carries forward into the post-combustion region of the incinerator, along with flyash. Inorganic chlorides (for example, sodium chloride or ferric chloride) are equally effective sources of chlorine.[36] M, the metal catalyst in the flyash, is typically copper (the most effective catalyst material[36]), though potassium, sodium, and zinc were also positively correlated with PCDD/F formation.[37] Homogeneous gas phase reactions in the combustion zone (see the Section on Homogeneous Gas Phase Reactions) generate a range of PICs, both chlorinated and non-chlorinated, which adsorb onto the surface of the flyash. Chlorine obtained from the Deacon reaction chlorinates the hydrocarbon species to add to the fund of chlorinated precursors on the flyash. These precursors react to form PCDDs and PCDFs. Taking dichlorobenzene and *o*-chlorophenol as typical precursors, the formation of monochloro-CDD can proceed initially by oxygen attack on dichlorobenzene to form a phenoxy radical, which reacts with chlorophenol to form an intermediate diphenyl ether. Combination of the hydrogen from the hydroxyl radical with chlorine on the adjacent phenyl ring

[37] W. S. Hinton and A. M. Lane, *Chemosphere*, 1991, **23**, 831.

allows the remaining oxygen to bridge between the two phenyl rings and complete the dioxin molecule.[38]

The catalytic formation of PCDDs and PCDFs has been studied by a number of workers.[35,36] Two competing, temperature-dependent reactions operate: formation of dioxin (reaching a maximum rate in the region of 300 °C) and dechlorination/decomposition of dioxin, a reaction whose rate increases exponentially with temperature. Below about 200 °C the rate of formation is very low. Between 200 °C and 400 °C the rate of formation exceeds the rate of destruction, peaking at about 300 °C. Above 400 °C the rate of destruction dominates, and dioxin levels decrease rapidly. A global heterogeneous mechanism has been proposed,[39,40] which has been shown to be in good agreement with experimental data. The model is in four stages:

(1) Dioxin formation: $P_g + P_s \rightarrow D_s$
(2) Dioxin desorption: $D_s \rightarrow D_g$
(3) Dechlorination: $D_s \rightarrow$ Products
(4) Decomposition: $D_s \rightarrow$ Products

P_g and P_s are precursor concentrations in the gas phase and on the surface of the flyash, and D_s and D_g are concentrations of dioxin on the flyash and in the gas phase. A proportion of the PCDD/Fs that are formed desorb off the flyash and exit the incinerator as a component of the gas phase. However, apart from the mono-, di-, and tri-chloro species, the major proportion of PCDD/Fs remain adsorbed onto flyash, and either exit the stack as a component of the particulate fraction, or are arrested in air pollution control equipment such as fabric filters.

Heterogeneous gas–solid reactions have also been implicated in the chlorination of organics such as naphthalene, biphenyl, and anthracene, under cooler conditions typical of the boiler and dust arrestment components of an incinerator.[41] Chlorinated products such as 2-, 3-, and 4-monochlorobiphenyl and 9,10-dichloroanthracene were identified, the overall yield approaching 10% of the starting material. Both the parent compounds and the products were strongly adsorbed onto flyash.

Heating of flyash in the presence of oxygen results in a ten-fold increase in the PCDD/F concentration relative to the untreated flyash.[35,36] Removal of oxygen from the reaction atmosphere followed by thermal treatment of the flyash results in a dramatic fall in PCDD and PCDF concentrations.[42] In addition to blocking the Deacon reaction, an oxygen-deficient environment encourages thermal decomposition of these compounds via catalysed dechlorination/hydrogenation reactions, again with copper as the most effective dechlorination agent. An appreciation of such reaction mechanisms assists in the development of control systems to minimize emissions of organic micropollutants, the subject of the next section.

[38] C. M. Young and K. J. Voorhees, *Chemosphere*, 1991, **23**, 1265.
[39] E. R. Altwicker, J. S. Schonberg, R. Konduri, and M. S. Milligan, *Haz. Waste Haz. Mater.*, 1990, 7, 73.
[40] R. Kolluri and E. R. Alteicker, *J. Air Waste Manage. Assoc.*, 1992, **42**, 1577.
[41] G. A. Eiceman and R. V. Hoffman, in 'Emissions from Combustion Processes: Origin, Measurement, Control', ed. R. Clement and R. Kagel, Lewis Publishers, Boca Raton, 1990, p. 71.
[42] H. Hagenmaier, H. Brunner, R. Haag, and M. Kraft, *Environ. Sci. Technol.*, 1987, **21**, 1085.

4 Control of Emissions

Given the toxicity of some of the organic micropollutants that can be formed in combustion systems, recent regulatory attention has focused on the need to minimize these emissions, and to control concentrations in the released gases to within stipulated limits. This section discusses some of the operational control and gas cleaning techniques that have been applied, and their rationale in terms of the combustion mechanisms introduced in previous sections.

Operational Control

Control of Feed Composition. Combustion conditions are best controlled when the system is not subjected to sudden changes in waste feed composition. Chemicals with different vapour pressures, combustibility, and heat content make different demands on the oxygen available in the incinerator—if waste feed characteristics are not matched to the incinerator operating conditions, overloading can occur relative to the amount of oxygen available for combustion. Rapid volatilization and combustion of waste components can cause localized depletion of oxygen levels, leading to less than optimum destruction of POHCs, and pyrolytic conditions conducive to the formation of PICs. The resulting transient releases or 'puffs' of higher concentrations of POHCs and PICs in the stack gas are often associated with excursions of CO or total hydrocarbon concentrations above their respective regulatory limits. Transient failures or 'upset' conditions are of particular concern in incinerators such as rotary kilns, accepting drums, or containers in batch form. A similar effect can occur in liquid waste incinerators if the feed is a variable mixture of waste types, typical of commercial incineration operations that accept materials from third parties.

Blending of the liquid wastes and ensuring uniformity in the introduction of solids and sludges is therefore vital if upset conditions are to be avoided. A blending strategy for liquids can be developed on the basis of the gasification behaviour of mixtures[43] whereby three types of waste are mixed: two groups that are readily incinerable but which have differing volatilities, and one group of wastes of intermediate volatility but demonstrating the greatest resistance to thermal destruction. The most volatile group gasifies first, and being readily incinerable, maintains a steady flame. The group of intermediate volatility gasifies next, the combustion of which is assisted by the gasification of the third, again readily incinerable group of wastes. Thus, flame conditions are optimized against the incinerability of the most refractory components of the waste feed. Atomization of liquid mixtures or slurries can be facilitated by blending waste components of similar concentrations but different volatilities,[43] the volatilization of the most volatile component causing violent rupture and disintegration of the droplet into the flame.

Regarding solid wastes, shredding followed by blending can assist in maintaining uniformity of combustion, as demonstrated in Figure 2.[44] The upper trace

[43] C. K. Law, in 'Incineration of Hazardous Waste—Toxic Combustion By-products', ed. W. R. Seeker and C. P. Koshland, Gordon and Breach Science Publishers, Philadelphia, 1992, p. 1.
[44] J. D. Kilgroe, L. P. Nelson, P. J. Schindler, and W. S. Lanier, in 'Incineration of Hazardous

Figure 2 The effect of
waste shredding on
emissions of carbon
monoxide

A. Single-stage shredding

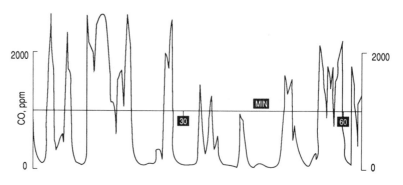

B. With secondary shredding

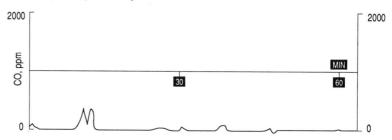

represents continuous measurements of CO in the stack gas while refuse derived
fuel (RDF) processed in a single-stage shredder is fed into a circulating fluidized
bed. Frequent transient excursions are evident over the measurement period of
one hour, indicative of perturbations in oxidative status. The lower trace
illustrates CO measurements when a two-stage shredder is used, resulting in a
finer and more uniform feed. Combustion conditions are now more stable,
eliminating the majority of transient peaks.

Control of Combustion Conditions. This includes control of temperature,
combustion air, and mixing within the incinerator. Thermal destruction
increases exponentially with temperature. Most regulatory systems require a
minimum destruction of 99.99% for POHCs within the waste feed, but 99.9999%
for PCBs. Minimum gas phase residence time and temperature criteria are also
required; typically 850 °C for MSW and sewage sludge, and 1100 °C to 1200 °C
for chlorinated chemical wastes, at a minimum of 2 s residence time in the
afterburner or secondary combustor, after the waste components have volatilized.
As indicated in Section 3, excess combustion air is maintained in the system to
ensure that oxidative conditions prevail. Intimate mixing of waste, auxiliary fuel,

Waste—Toxic Combustion By-products', ed. W. R. Seeker and C. P. Koshland, Gordon and
Breach Science Publishers, Philadelphia, 1992.

Table 5 Emissions of organic micropollutants at different MSW loading rates, [16] (μg Nm^{-3})

	Low	Burning rate Design	High
Chlorobenzenes	3.5	3.3	4.4
Chlorophenols	9.5	5.1	8.0
PAHs	7.1	4.0	5.4
PCBs	4.3	3.0	4.9
PCDDs	0.0526	0.0188	0.0554
PCDFs	0.1145	0.0445	0.1007

and air is an essential requirement for good oxidative combustion, and to prevent the development of localized fuel-lean, fuel-rich, or oxygen-deficient pockets in which PICs can form. Apart from the design of the combustion chamber and burners, the loading rate of waste into the incinerator controls the temperature and oxygen status within the combustor. At very low feed rates, combustion air requirements may be insufficient to maintain high temperatures or adequate turbulence within the system. At high feed rates gas production increases, reducing the residence time, and encouraging channelling of volatilized components through the chamber. Excess oxygen levels may be lowered, resulting in the release of transient puffs of PICs. Table 5 presents data obtained under different loading conditions, from the NITEP trial burns (see Section 2).[16] At both low and high loading rates relative to the design capacity of the plant, excessive concentrations of organic micropollutants were released.

Maintaining good combustion conditions has important implications for micropollutant control downstream of the combustor. The reaction mechanisms for PCDD/F formation in the boiler and dust arrestment components of the plant are dependent upon the availability of an organic substrate adsorbed onto particulate matter exiting the combustion zone. More complete destruction of the waste components results in less carryover of unburnt organics and PICs, and therefore a reduced potential for heterogeneous reactions involving precursors of PCDDs, PCDFs, and PAHs.

Control of Surrogate Emissions. Since continuous measurement of individual organic micropollutant concentrations in the stack gas is as yet unattainable, other indicators of good combustion are generally monitored on a continuous basis, along with batch testing for specific organics such as PCBs and PCDD/Fs. These indicators include temperature and excess oxygen (for the reasons discussed above), and CO as a surrogate for organic emissions. In-stack instrumental analysis of CO can be readily undertaken, and serves as a sensitive real-time indicator of poor combustion conditions, as is evident in Figure 3. Total hydrocarbon (THC) emissions are often measured alongside CO emissions, but kinetic considerations favour the earlier release of CO under oxygen-deficient conditions; only when combustion deteriorates to a significant extent do THC emissions increase appreciably. An emission limit on CO concentration now forms an integral part of all regulatory systems.

While an increase in CO emissions provides a general indication of deteriorating combustion conditions, a low CO emission concentration does not necessarily

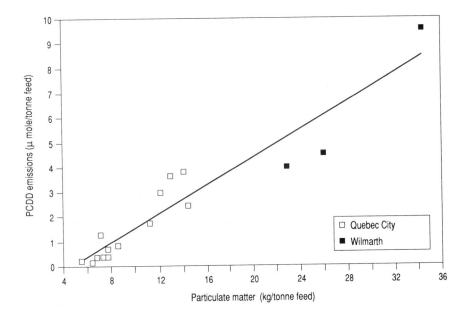

Figure 3 Particulate
arrestment and emissions
of PCDDs

imply low PCDD/F or PAH emissions. Post-combustion heterogeneous reactions in downstream equipment may result in an increase in PCDD/F formation, while experiments on PAH formation in chlorine environments have indicated the lack of a correlation between CO and PAH emissions.[45,46] Separate monitoring for these chemicals is therefore necessary to confirm the acceptability of their emissions.

Control of Post-combustion Conditions

Control of Post-combustion Temperature. The gases from the combustion zone must be cooled from 850 °C–1200 °C before entering the air pollution control component of the incineration system. Since most pollution control devices operate below 300 °C, the gases must pass through the window of 200 °C–400 °C, within which PCDD/F formation is of concern. Rapid cooling of the gases can be achieved by quenching with water. Alternatively, if the plant is designed for heat recovery, then maintaining the boiler outlet temperature in the region of 350 °C minimizes the potential for PCDD/F formation; additional cooling, if required, is achieved through the introduction of air. More frequent rapping and cleaning of the boiler tubes and dust arrestment equipment prevents build-up of deposits on hot surfaces, ensuring that the flyash is not subjected to the temperature range of concern for long periods.

Air Pollution Control Devices. Another strategy to minimize emissions of organic micropollutants is to remove them from the combustion gases downstream

[45] L. J. Staley, M. K. Richards, G. L. Huffman, R. A. Olexsey, and B. Dellinger, *J. Air Pollut. Control Assoc.*, 1989, **39**, 321.

[46] M. Frenklach, in 'Incineration of Hazardous Waste—Toxic Combustion By-products', ed. W. R. Seeker and C. P. Koshland, Gordon and Breach Science Publishers, Philadelphia, 1992, p. 283.

of the boiler or quench. The knowledge that PAH, PCDD, PCDF, and other organic micropollutants adsorb strongly onto particulate matter offers a means of control. Figure 3[47] indicates a direct relationship between the efficacy of particulate arrestment and removal from the gas stream, and a reduction in PCDD/F emissions. Fabric filters are increasingly used for this purpose, since they operate to higher particulate removal efficiencies at lower particle sizes—although a minor fraction of the total size range, these flyash particles present a greater surface area for adsorption and reaction, and hence have the potential to transport a larger proportion of the organic micropollutants to the stack and into the atmosphere.

Injection of activated carbon into the fabric filter is another effective means of improving the removal of organic micropollutants from the gas stream, in particular light hydrocarbons and lower-chlorinated PCDDs and PCDFs that are predominantly in vapour form rather than adsorbed onto particulates. The build-up of a layer of flyash and carbon on the fabric ensures intimate contact between the gases and the surface of the particles, as the gas stream is drawn through the filter.

A novel catalytic reactor has recently been installed on an MSW incinerator, comprising a conventional de-NO_x catalyst base of titanium dioxide doped with tungsten and vanadium oxides, and coated with additional proprietary materials.[48] Pilot plant and demonstration scale tests have indicated effective oxidation of PCDDs and PCDFs to carbon dioxide, HCl, and water, with the simultaneous catalytic reduction of nitrogen oxides to nitrogen, by injection of ammonia.

Treatment of Flyash.   A consequence of removing organic micropollutants from the gas stream is the accumulation of these chemicals in the flyash/carbon deposits collected from the boilers, fabric filters, and other dust arrestment points. The disposal of this material in landfills has been subject to increasingly stringent controls, leading to the search for techniques to detoxify the ash so as to ease these problems. Vitrification or solidification of the flyash has been proposed, to reduce the organic content of the material, and the leachability of metals. A benefit of this form of post-treatment is that the resulting material can be utilized as low-grade aggregate for road works and landscaping, or in the case of vitrification, as a cheap glass with industrial sealing applications.

In Section 3 the catalytic dechlorination/hydrogenation of PCDD/Fs on flyash was discussed. This reaction mechanism, resulting in a significant decrease in PCDD/F concentrations, is the basis of a commercial thermal detoxification process.[48,49] Flyash is collected from the incineration plant and transferred to a vessel maintained at 400 °C and less than 1% oxygen. The flyash contains sufficient amounts of metals, such as copper, to obviate the need for the addition of a catalyst. Trials on full-scale plant have shown reductions in PCDD/F levels in the flyash, of between 95–99%.

[47] T. G. Brna and J. D. Kilgroe, *J. Air Waste Manage. Assoc.*, 1990, **40**, 1342.
[48] G. Schleger, Proceedings of the Conference on Incineration—The Great Debate, IBC Technical Services Limited, Manchester, 1992.
[49] H. Hagenmaier, K.-H. Tichaczek, M. Kraft, R. Haag, and H. Brunner, European Patent No. 0 252 521, 1987.

5 Health Effects

Public concern and research effort on the potential adverse health effects of emissions of organic micropollutants has focused on a few highly publicized chemicals, in particular PCBs, PCDDs, and PCDFs, to the exclusion of almost all other chemical types. Health risk assessments on incinerators operating to modern standards have belied the emphasis placed on these chemicals; the overall health risk is more often than not driven by emissions of inorganic species than by emissions of the organics. This section discusses the magnitude of potential health effects from organic emissions, and presents the results of a typical risk assessment on an incinerator operating to current emission standards for organic micropollutants.

Health Effects and Exposure Pathways

Organic micropollutants can be divided into two broad categories: non-carcinogens and carcinogens. The latter category can be further divided into two groups, *genotoxic* carcinogens that initiate cancer through an initial effect on DNA or chromosomes, and *non-genotoxic* carcinogens that inflict chronic cell damage. The dose–effect relationship of non-carcinogens and non-genotoxic carcinogens is generally assumed to be bounded at low doses by a threshold, below which these chemicals fail to induce a discernible adverse health effect. This threshold is termed the reference dose (RfD) in the US, and the Acceptable Daily Intake (ADI) or Tolerable Daily Intake (TDI) in Europe, and is set by applying a safety factor to doses which are known not to elicit observed adverse effects (the so-called No Observed Effect Level, or NOEL). For genotoxic carcinogens it is generally assumed that there is no such threshold. In the US, probabilistic risk assessment is used to estimate the health risk resulting from exposure to the genotoxic chemical: a carcinogenic potency factor called the *slope factor* is computed from the dose–response data, which, when multiplied with the calculated intake of the carcinogen, provides an estimate of the likelihood of mortality from cancer. In Europe the safety factor approach is applied, to generate an ADI or TDI for a genotoxic carcinogen similar to that derived for a non-carcinogen or non-genotoxic carcinogen.

Exposure to organic micropollutants emitted into the atmosphere via the stack can be *direct*, for example through inhalation, deposition onto the skin, and ingestion or contact with soil contaminated by the emissions; or *indirect*, for example through the consumption of fruits and vegetables contaminated with chemicals depositing onto plant surfaces, or through the consumption of meat and dairy products derived from animals reared on contaminated feed. Figure 4 provides a schematic representation of some of the various exposure pathways to humans, following the release of organic micropollutants.[50]

The physicochemical properties of the organic chemical will dictate the relative prominence of one or other of these exposure routes: for low molecular weight, volatile organics the dominant exposure pathway is likely to be inhalation,

---

[50] J. Petts and G. H. Eduljee, 'Environmental Impact Assessment for Waste Treatment and Disposal Facilities', John Wiley, Chichester, 1994.

Typical exposure
pathways to humans,
release of organic
micropollutants to
atmosphere

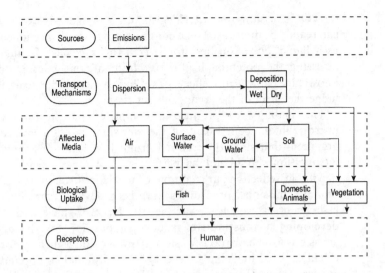

whereas for compounds such as PCDDs and PCDFs, indirect pathways are of greater importance. Thus, for trichloroethylene, inhalation, ingestion of vegetables, and ingestion of dairy products contribute 56%, 27%, and 6% respectively to the total daily uptake of this chemical,[51] whereas for 2,3,7,8-TCDD the corresponding contributions are 3%, 16%, and 36% respectively.[52]

The potential for exposure will clearly be site-specific. The extent to which the public will come into contact with emissions or contaminated environmental media such as soil will depend on the activity patterns in the vicinity of the incinerator. For example, the absence of significant agricultural activity will lessen the contribution of this pathway; if the tending of allotments is practised, then uptake via vegetables and fruits may be of significance, as opposed to uptake via dairy products.

Risk Assessment

Regulators controlling emissions of organic micropollutants generally bulk all such emissions into one of two categories: total hydrocarbons (THCs) or total organic carbon (TOC). Emission limits are then assigned to these groups—for example, 20 p.p.m.v. for THC emissions under US legislation,[53] and $20\,mg\,m^{-3}$

[51] T. E. McKone and P. B. Ryan, 'Human Exposure to Chemicals through Food Chains: An Uncertainty Analysis', Report No. UCRL-99290, Lawrence Livermore National Laboratory, University of California, 1989.

[52] ECETOC, 'Exposure of Man to Dioxins: A Perspective on Industrial Waste Incineration', Technical Report No. 49. European Centre for Ecotoxicology and Toxicology of Chemicals, Brussels, 1992.

[53] US EPA, 'Burning of Hazardous Waste in Boilers and Industrial Furnaces; Final Rule', Federal Register, 56, 7134, February 21, 1991, and 56, 42504, August 27, 1991.

for TOC emissions under UK legislation.[54] While THCs or TOC in the stack gas can readily be monitored on a continuous basis with instrumental analysers such as a flame ionization detector (FID), this approach is unhelpful in terms of assessing the potential health risk from organic micropollutants; without a knowledge of the health effects posed by individual chemicals, it is not possible to judge the effect of the composite emissions.

The inability to produce a complete characterization of the organic micropollutant component of the emissions is often invoked to support the argument that since the true risk from these emissions cannot be determined, incineration should not be permitted.[55] While an element of uncertainty is inevitable, reflecting our incomplete knowledge of all the chemicals present in the organic emissions, procedures have been devised that allow for a worst-case assessment of risk. For example, the US EPA have addressed this issue by developing a conservative generic composition for the organic fraction in emissions from hazardous waste and sewage sludge incinerators by adding to the organics listed in Table 1, seventy-five additional compounds of toxicological concern selected from the US EPA's Appendix VIII. To compensate for uncertainties in their concentrations, each additional compound was conservatively included at the limit of detection, *i.e.* $100 \, \mu g \, m^{-3}$. A weighted slope factor for the carcinogenic fraction of the total emissions was then calculated, permitting an evaluation of the carcinogenic risk posed to an exposed individual.[19,20]

A calculation can be made of the inhalation risk posed by corporate emissions of organic micropollutants. The weighted average molecular weight of the generic emissions ranges from $45 \, g \, mol^{-1}$ for hazardous waste incineration to $34 \, g \, mol^{-1}$ for sewage sludge incineration,[19,20] indicating that low molecular weight compounds dominate the emissions. Assuming an average carbon content of 50%, a TOC emission limit of $20 \, mg \, Nm^{-3}$ is equivalent to an emission concentration for THCs of $40 \, mg \, Nm^{-3}$, or an emission rate of $200 \, mg \, s^{-1}$ for an MSW incinerator with a capacity of 400 000 tonnes year^{-1}.[56] This emission rate results in a maximum annual average ambient air ground-level concentration of $0.04 \, \mu g \, m^{-3}$, which translates into a conservative lifetime carcinogenic risk of 1 in 55 million to an individual located at the point of maximum impact of the emissions. Even if indirect exposure pathways were added, the total conservative lifetime carcinogenic risk for a maximally exposed individual would still fall well below 1 in 1 million, with more realistic exposure assumptions reducing this further. This level of risk is below involuntary risks such as, say, death from lightning[56] and also below typical regulatory criteria for acceptable risk of 1 in 100 000 to 1 in 1 000 000. A similar conclusion was reached via another approach,[17] in which the combined emission of organic micropollutants detected in the stack gases of hazardous waste incinerators was multiplied by a factor of 100 to compensate for a 1% level of detection, and the resulting emission

[54] Her Majesty's Inspectorate of Pollution, 'Environmental Protection Act 1990—Process Guidance Note IPR 5/1', HMSO, London, 1992.

[55] Greenpeace, 'Proof of Evidence to the local Inquiry held to hear the Appeal by Cory Environmental Management Limited for Land at Seal Sands, Billingham', Application No. CS/2262/88, Department of the Environment, Northern Regional Office, Newcastle upon Tyne, 1990.

[56] Environmental Resources Management (ERM), unpublished results.

rate treated as discussed above to arrive at an estimate for the conservative carcinogenic risk experienced by a maximally exposed individual. It should be noted that PCDDs and PCDFs are included in the US EPA generic organic micropollutant composition, and therefore their potential effect on human health has been accounted for in the calculation of carcinogenic risk.

Consideration of non-carcinogenic health effects involves the selection or derivation of an appropriate air quality standard or ADI for the chemical of concern, against which the calculated ambient air ground-level concentration or calculated uptake through the foodchain can be compared. An assessment of potential non-carcinogenic effects due to emissions of chlorobenzenes, chlorophenols, phthalates, and PCBs from the stack of a proposed 1.2 million tonnes year^{-1} MSW incinerator indicated that the calculated ground-level concentrations of these chemicals at the point of maximum impact fell below their respective air quality standards by factors of 1 million to 10 million,[57] signifying an ample margin of safety from the standpoint of providing protection against an adverse health effect.

6 Conclusions

An understanding of the mechanisms of formation of POHC and PIC emissions is essential for the design of control systems, and for setting appropriate operating conditions which optimize oxidative combustion. This is perhaps best illustrated in the case of PCDDs and PCDFs, where an appreciation of the significance of post-combustion reformation reactions has resulted in new strategies to minimize and control these emissions. Current advice on good combustion practice is based on a relatively small body of research and engineering experience, and there is considerable scope for further improvements in combustor design and control in order to limit the formation of organic micropollutants. A particular challenge is to devise blending and control strategies that smooth out the transient upset conditions discussed in Section 4.

Regulation of organic micropollutant emissions from incinerators is achieved through a combination of restrictions on operating variables (temperature, gas phase residence time, and excess oxygen requirements) and adherence to emission limits. While monitoring of emissions of surrogate compounds such as CO provides an indication of the global destruction efficiency of POHCs in the waste feed and an indirect measure of the propensity for PIC formation, continuous on-line measurement of the micropollutants of greatest concern (PCBs, PCDDs, PCDFs, *etc.*) is recognized by both regulators and operators of incineration facilities as essential to address public anxiety over adverse health effects, and to reduce the uncertainties in public health risk assessments. Promising research in advanced instrumental analysis is being undertaken,[58] but these systems have yet to be commercialized.

---

[57] J. W. Bridges, 'Proof of Evidence in Support of the Application by Cory Environmental Limited to Construct and Operate a Refuse to Energy Plant at Belvedere, Bexley', Inquiry under the Electricity Act, Ref. B257/P53/1, Department of Energy, London, 1992.

[58] E. A. Rohlfing, D. W. Chandler, and G. A. Fisk, Proceedings of the AFRC Symposium on Incineration of Municipal, Hazardous, and Other Wastes, Palm Springs, California, 1987.

Despite the non-availability of a complete characterization of the organic fraction of stack gas emissions, it is possible to apply risk assessment techniques to estimate the health risk posed to potentially exposed members of the general population. Compensating for uncertainties in micropollutant composition and concentrations by incorporating conservative assumptions on emission rates and exposure potential, it can be concluded that an incinerator operating to current emission limits for organic micropollutants should pose a negligible health risk to the surrounding community, though this does not negate the fundamental importance of siting such plant in locations that minimize the potential for exposure to stack emissions.

Pilot-scale Research on the Fate of Trace Metals in Incineration

G. J. CARROLL

1 Introduction

Since the mid-1970s, the US Environmental Protection Agency (EPA) Office of Research and Development (ORD) has studied the incineration of hazardous waste. Early studies focused largely on the destruction and removal efficiencies of incinerators with respect to hazardous organic compounds and on the particulate removal efficiencies of air pollution control systems.[1-4] In more recent years, attention has turned to the potential risks associated with stack emissions of carcinogenic and toxic metals from such incinerators.

Bench- and pilot-scale research studies on metal partitioning (the distribution of metals among incinerator discharges) have been conducted and sponsored by the Combustion Research Branch of ORD's Air and Energy Engineering Research Laboratory (AEERL–RTP, NC)[5-9] and the Thermal Destruction Branch of ORD's Risk Reduction Engineering Laboratory (RREL—Cincinnati, OH). In addition, pilot- and full-scale studies on the subject have been

[1] US EPA, 'Determination of Incinerator Operating Conditions Necessary for Safe Disposal of Pesticides', July 1975.
[2] US EPA, 'Emission Test Results for a Hazardous Waste Incineration Regulatory Impact Analysis', Midwest Research Institute, EPA/600/9-84/015, 1984.
[3] US EPA, 'Performance Evaluation of Full-scale Hazardous Waste Incinerators', 5 Volumes, NTIS, PB 85-129500, Nov. 1984.
[4] US EPA, 'Assessment of Incineration as a Treatment Methodology for Liquid Organic Hazardous Wastes—Background Report V', March 1985, US GPO 1985-526-778/30376.
[5] J. A. Mulholland, G. Yue, and A. F. Sarofim, 'The Formation of Inorganic Particles During Suspension Heating of Simulated Wastes', *Environ. Prog.*, Aug. 1990.
[6] V. Sethi and P. Biswas, 'Modeling of Particle Formation and Dynamics in a Flame Incinerator', *J. Air Waste Manage. Assoc.*, 1990, **40**, 42.
[7] M. V. Scotto, T. W. Peterson, and J. O. Wendt, 'Hazardous Waste Incineration: The *In-situ* Capture of Lead by Sorbents in a Laboratory Down-flow Combustor', Twenty Fourth Symposium (International) on Combustion, The Combustion Institute, 1992.
[8] W. P. Linak, R. K. Srivastava, and J. O. Wendt, 'Metal Aerosol Formation in a Laboratory Swirl Flame Incinerator', submitted to *Combust. Sci. Technol.*, 1993.
[9] W. P. Linak and J. O. Wendt, 'Toxic Metal Emissions from Incineration: Mechanism and Control', *Prog. Energy Combust. Sci.*, 1993.

95

G. J. Carroll

sponsored by EPA's Office of Solid Waste (OSW—Washington, DC).[10-12]

This article examines RREL pilot-scale research conducted at EPA's Incineration Research Facility (IRF) in Jefferson, AR. It is intended to serve as an overview and is based on a sample of the IRF studies in which metal partitioning was investigated.

2 Background

Regulation of Metal Emissions

In January 1981, regulations governing the incineration of hazardous waste were promulgated by EPA pursuant to the Resource Conservation and Recovery Act (RCRA).[13] Among the performance standards applied to hazardous waste incinerators was a stack gas limit on particulate emissions of 180 mg dscm^{-1}*. Until recently, that standard was the only means, albeit indirect, by which stack emissions of metals were regulated.

In April 1990, EPA proposed amendments to the hazardous waste incinerator regulations to provide improved control of toxic metal emissions, hydrogen chloride emissions, and residual organic emissions.[14] Although the amendments have not yet been adopted, much of their substance has been put into practice by regulatory authorities via a permit provision known as *omnibus* authority (a provision allowing regulatory authorities to add to permits 'terms and conditions as the Administrator or State Director determines necessary to protect human health and the environment').

EPA's three-tiered approach to control of metal emissions is structured to allow higher emission rates (and feed rates) of metals as incinerator owners/operators elect to conduct more site-specific testing and analysis. Any one or a combination of the three approaches/tiers may be used.

The controls are based on the projected inhalation risks established for ten toxic metals and incorporate the ambient levels of the metals that EPA believes pose acceptable health risk. Under Tier I, the simplest but most conservative approach, limits are set on the hourly feed rate of metals to the incinerator. These feed rates are back-calculated from acceptable ambient air quality levels using conservative air dispersion modeling and assuming that 100% of the metals in the feed are emitted at the stack (no partitioning of metals to the bottom ash occurs, nor are metals removed by the air pollution control system [APCS]).

The Tier II approach limits the stack emission rates of metals. As with Tier I,

---

* dscm = dry standard cubic metre.

10 US EPA, 'Measurement of Particulates, Metals, and Organics at a Hazardous Waste Incinerator', Midwest Research Institute, NTIS No. PB89-230668, Nov. 1988.
11 US EPA, 'Pilot-Scale ESP and Scrubber Parametric Tests for Particulates, Metals, and HCl Emissions: John Zink Company Research Facility; Tulsa, OK', Radian Corp., NTIS No. PB90-129362, June 1989.
12 US EPA, 'A Performance Test on a Spray Drier, Fabric Filter and Wet Scrubber System: APTUS Incinerator at Coffeyville, KS', Radian Corp, NTIS No. PB90-120544, Oct. 1989.
13 US EPA, 'Incinerator Standards for Owners and Operators of Hazardous Waste Management Facilities: Interim Final Rule and Proposed Rule'; *Fed. Reg.*, 1981, **46** (No. 15), 7666-7690.
14 US EPA, 'Standards for Owners and Operators of Hazardous Waste Incinerators . . .', *Fed. Reg.*, 1990, **55** (No. 82), 17862-17921.

emission rates are established by back-calculating from acceptable ambient air levels using conservative air dispersion modeling assumptions. Tier II differs from Tier I, however, in that emissions testing enables the owner/operator to take credit for reduced metal emissions achieved either by partitioning of metals to the bottom-ash, or by removal of metals by the APCS.

Site-specific emissions can be determined under Tier III by performing site-specific air dispersion modeling. As with Tier II, compliance with back-calculated emission limits is confirmed by stack sampling.[15,16]

Data Needs

In 1987, E. Timothy Oppelt, now Director of RREL, examined the state of knowledge regarding hazardous waste incineration. Among his findings was the fact that, while the human health risk of incinerator emissions appeared to be low, metal emissions were the dominant component of risk levels identified. Oppelt also found that insufficient data existed relative to the following: physical and chemical characteristics of particulate matter; particle size distribution of emissions; metal removal capabilities of APCSs; and the fate/partitioning of metals among incinerator discharges.[17]

Incineration Research Facility (IRF)

The IRF is an experimental facility that currently houses pilot-scale incineration systems and their associated waste handling, emission control, process control, and safety equipment. The IRF also has on-site laboratory facilities for waste characterization and analysis of process performance samples. Among the objectives of the research projects conducted at the IRF are the following:

(i) *to develop incinerator system performance data* for regulated hazardous wastes to support RCRA incinerator regulations and performance standards, and to provide a sound technical basis for any future standards;

(ii) *to promote an understanding of the hazardous waste incineration process* and develop methods to predict the performance of incinerators of varying scale and design as a function of key process operating variables;

(iii) *to test the performance of new and advanced incinerator components, subsystems, and pollution control devices*; and

(iv) *to provide a means for conducting specialized test burns* (particularly for high-hazard or special waste materials such as Superfund site wastes) in support of Regional Office permitting or enforcement actions and Regional Office or private-party Superfund site remediation efforts.[18]

Pursuant to these objectives, and in recognition of the aforementioned data

[15] C. R. Dempsey and E. T. Oppelt, 'Incineration of Hazardous Waste: A Critical Review Update', *J. Air Waste Manage. Assoc.*, 1993, **43**, 25.

[16] US EPA, 'Guidance on Metals and Hydrogen Chloride Controls for Hazardous Waste Incinerators: Volume IV of the Hazardous Waste Incineration Guidance Series', Aug. 1989.

[17] E. T. Oppelt, 'Incineration of Hazardous Waste: A Critical Review', *JAPCA*, 1987, **37**, 558–586.

[18] US EPA, 'Operations and Research at the US EPA Incineration Research Facility: Annual Report for FY92', Acurex Environmental Corporation, June 1993, EPA/600/R-93/087.

Figure 1 IRF rotary kiln
system schematic

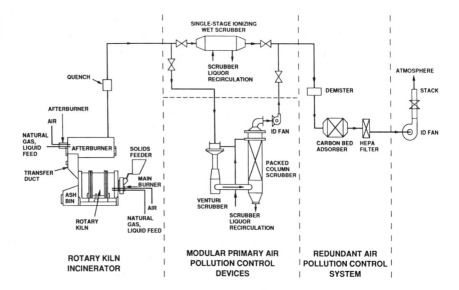

needs, a number of research studies have taken place at the IRF over the past several years in which the partitioning of metals has been addressed. These studies are described herein.

3 Test System

All of the subject tests took place in the pilot-scale rotary kiln system (RKS) at the IRF. The RKS, depicted in Figure 1, consists of a rotary kiln primary chamber followed by a transition duct and afterburner chamber. Combustion gases exiting the afterburner are quenched, after which they enter a primary APCS. This is followed by secondary/backup APCS consisting of a demister, carbon bed adsorber, and high efficiency particulate filter.

The RKS configuration provides several options for primary acid-gas and particulate control. The most frequently used APCS is a venturi/packed-column scrubber combination. Also available for use with the RKS is a single-stage ionizing wet scrubber.

A third option in the RKS design allows for in-line evaluation of modular, pilot-scale APCSs. One such system, a Calvert Flux-Force/Condensation Scrubber, was in place during two of the IRF test series described below. The main components of the Calvert system were a condenser/absorber; the Collision Scrubber (which splits the flue gas into two streams, then uses head-on collision to create fine droplets and a large surface area to enhance downstream particulate and acid-gas removal); an entrainment separator; a wet electrostatic precipitator; another entrainment separator; and a variable-speed induced-draft fan.[19]

Nominal design characteristics of the RKS components may be found in Table 1.

[19] US EPA, 'The Fate of Trace Metals in a Rotary Kiln Incinerator with a Calvert Flux Force/Condensation Scrubber' (2 Volumes)—Draft Report, Acurex Environmental Corp., Jan. 1993.

Table 1 Design characteristics of the IRF rotary kiln system

Characteristics of the Kiln Main Chamber

Length	2.49 m (8 ft-2 in)
Diameter, outside	1.37 m (4 ft-6 in)
Diameter, inside	Nominal 1.00 m (3 ft-3.5 in)
Chamber volume	1.90 m^3 (67.3 ft^3)
Construction	0.95 cm (0.375 in) thick cold-rolled steel
Refractory	18.7 cm (7.375 in) thick high alumina castable refractory, variable depth to produce a frustroconical effect for moving solids
Rotation	Clockwise or counterclockwise, 0.2 to 1.5 r.p.m.
Solids retention time	1 h (at 0.2 r.p.m.)
Burner	North American burner rated at 590 kW (2.0 MMBtu h^{-1}) with liquid feed capability
Primary fuel	Natural gas
Feed system:	
Liquids	Positive displacement pump via water-cooled lance
Sludges	Moyno pump via front face, water-cooled lance
Solids	Metered twin-auger screw feeder or fiberpack ram feeder
Temperature (max.)	1010 °C (1850 °F)

Characteristics of the Afterburner Chamber

Length	3.05 m (10 ft)
Diameter, outside	1.22 m (4 ft)
Diameter, inside	0.91 m (3 ft)
Chamber volume	1.80 m^3 (63.6 ft^3)
Construction	0.63 cm (0.25 in) thick cold-rolled steel
Refractory	15.2 cm (6 in) thick high alumina castable refractory
Gas residence time	1.2 to 2.5 s depending on temperature and excess air
Burner	North American Burner rated at 590 kW (2.0 MMBtu h^{-1}) with liquid feed capability
Primary fuel	Natural gas
Temperature (max.)	1200 °C (2200 °F)

Characteristics of the Ionizing Wet Scrubber APCS

System capacity, inlet gas flow	85 m^3 min^{-1} (3000 a.c.f.m.) at 78 °C (172 °F) and 101 kPa (14.7 p.s.i.a.)
Pressure drop	1.5 kPa (6 in WC)
Liquid flow	15.1 L min^{-1} (4 g.p.m.) at 345 kPa (50 p.s.i.g.)
pH control	Feedback control by NaOH solution addition

Characteristics of the Venturi/Packed-column Scrubber APCS

System capacity, inlet gas flow	107 m^3 min^{-1} (3773 a.c.f.m.) at 1200 °C (2200 °F) and 101 kPa (14.7 p.s.i.a.)
Pressure drop	
Venturi scrubber	7.5 kPa (30 in WC)
Packed column	1.0 kPa (4 in WC)
Liquid flow	
Venturi scrubber	77.2 L min^{-1} (20.4 g.p.m.) at 60 kPa (10 p.s.i.g.)
Packed column	116 L min^{-1} (30.6 g.p.m.) at 69 kPa (10 p.s.i.g.)
pH control	Feedback control by NaOH solution addition

4 Test Descriptions

For the purpose of this discussion, metal partitioning studies at the IRF may be grouped into one of two categories: *'fundamental'*, parametric research studies using synthetic, formulated wastes; and *'applied'* treatability studies using real-world, Superfund-site wastes. While test conditions were varied in both types of studies, the parametric studies were more extensive in their evaluation of operating conditions and waste feed composition as variables; the primary focus of the treatability studies was on the ability to incinerate the wastes in compliance with performance standards.

Parametric Test Programs

Three series of parametric metal partitioning studies have taken place at the IRF over the past several years. In 1988, an eight-test series examined metal behavior in the RKS using the venturi/packed-column scrubber as the primary APCS.[20] Nine tests in 1989 evaluated metal partitioning using the single-stage ionizing wet scrubber.[21] More recently, in 1991, the Calvert Flux-force/Condensation Scrubber was used in an eleven-test series.[19]

The waste feed for each of the parametric programs contained a mixture of organic liquids added to a clay absorbent material. Trace metals were incorporated by spiking aqueous mixtures of the following hazardous and non-hazardous metals onto the organic/clay material: arsenic, barium, bismuth, cadmium, chromium, copper, lead, magnesium, and strontium (non-hazardous metals were included to investigate whether their discharge distributions paralleled those of any of the hazardous metals in the mixture and to provide data to support the development of a numerical metal partitioning model). In the test program with the Calvert scrubber, mercury was also added to the test mixture.

Test variables in the first two programs were kiln exit gas temperature, afterburner temperature, and feed chlorine content (chlorine content was adjusted by varying the mixture of organic liquids in the synthetic waste). In the Calvert test program, scrubber pressure drop replaced afterburner temperature as a test variable.

The discussion of parametric test results in this article focuses on the venturi/packed-column and single-stage ionizing wet scrubber test series; results from the Calvert test program are under review.

Superfund Treatability Test Programs

Over the past several years the IRF has been solicited by EPA Regional Offices to conduct a number of treatability tests of wastes from Superfund sites. Among the sites for which studies were performed are: Baird and McGuire[22], New Bedford

---

[20] US EPA, 'The Fate of Trace Metals in a Rotary Kiln Incinerator with a Venturi/Packed-column Scrubber' (2 Volumes), Acurex Environmental Corp., EPA/600/R-90/043, Feb. 1991.

[21] US EPA, 'The Fate of Trace Metals in a Rotary Kiln Incinerator with a Single-Stage Ionizing Wet Scrubber' (2 Volumes), Acurex Environmental Corp., EPA/600/R-91/032, Sept. 1991.

[22] US EPA, 'Pilot-scale Incineration of Arsenic-contaminated Soil from the Baird and McGuire Superfund Site', Acurex Environmental Corp., May 1990.

Harbor[23], Chemical Insecticide Corporation[24], Drake Chemical[25], and Scientific Chemical Processing.[26]

Treatability studies were comprised of four to six tests in which some or all of the following parameters were addressed: decontamination effectiveness (ash *vs.* feed) for organic compounds, destruction and removal efficiencies (stack gas *vs.* feed) for organic compounds, metal partitioning, metal leachability (ash *vs.* feed); and emissions of particulate and hydrogen chloride. To a limited extent, the effects of operating conditions on these parameters were also explored. In some tests, metal partitioning objectives were secondary to investigating the fate of organic compounds (*e.g.* PCB destruction).

Site contaminants are briefly described below.

Baird and McGuire. Soils at the Baird and McGuire site (Holbrook, MS) are contaminated with pesticides at concentrations up to $1500 \, \text{mg} \, \text{kg}^{-1}$. Lead and arsenic are also found throughout the site at concentrations generally below $100 \, \text{mg} \, \text{kg}^{-1}$, but with hot spots for arsenic as high as $3800 \, \text{mg} \, \text{kg}^{-1}$.

New Bedford Harbor. Marine sediments from the hot spot in New Bedford Harbor (New Bedford, ME) are contaminated with PCBs at concentrations of 4000 to over $200\,000 \, \text{mg} \, \text{kg}^{-1}$, as well as with cadmium, chromium, copper, and lead at concentrations up to several hundred $\text{mg} \, \text{kg}^{-1}$.

Chemical Insecticide Corporation. Soils at the Chemical Insecticide Corporation (CIC) site (Edison Township, NJ) are highly contaminated with organochlorine pesticides (*e.g.* chlordane, DDT) and trace metals (*e.g.* arsenic at levels up to $8000 \, \text{mg} \, \text{kg}^{-1}$). Barium, cadmium, chromium, and lead are also present in the soils.

Drake Chemical. Soils at the Drake Chemical Site (Lock Haven, PA) are contaminated with varying degrees of volatile and semivolatile organics (*e.g.* Fenac, 2-butanone) and trace metals (arsenic, barium, chromium, copper, lead) at concentrations up to several hundred $\text{mg} \, \text{kg}^{-1}$.

Scientific Chemical Processing. Soils at the Scientific Chemical Processing (SCP) site (Carlstadt, NJ) are contaminated with a wide variety of volatile and semivolatile organics (including PCBs) and trace metals (barium, cadmium, chromium, copper, lead) at concentrations up to thousands of $\text{mg} \, \text{kg}^{-1}$.

5 Test Conditions

Data collected throughout the test programs included feed material composition, incinerator process variables, and discharge stream analysis results. The

[23] US EPA, 'Pilot-scale Incineration of PCB-contaminated Sediments from the New Bedford Harbor Superfund Site', Acurex Environmental Corp., EPA/600/R-92/069, Sept. 1992.

[24] US EPA, 'Pilot-scale Incineration of Contaminated Soil from the Chemical Insecticide Corporation Superfund Site'—Draft Final Report, Acurex Environmental Corp., Feb. 1993.

[25] US EPA, 'Pilot-scale Incineration of Contaminated Soil from the Drake Chemical Superfund Site', Acurex Environmental Corp., Feb. 1993.

[26] US EPA, 'Pilot-scale Incineration of Contaminated Soil from the Scientific Chemical Processing Superfund Site'—Draft Final Report, Acurex Environmental Corp., May 1993.

sampling and analysis protocol included measurements necessary to identify the distribution of metals among the three RKS discharge streams (kiln ash, scrubber liquor, and stack gas).

Table 2 reflects key conditions for the subject tests. In all but two of the Superfund treatability tests, the venturi/packed-column scrubber served as the primary APCS. The Calvert Scrubber was used during the CIC tests; the single-stage ionizing wet scrubber was used during the Baird–McGuire tests.

Waste feed metal concentrations may be found in Table 3.

6 Results

When subjected to incineration conditions, metals are expected to vaporize to varying degrees, depending on their relative volatilities. To predict a metal's volatility, equilibrium analyses can be performed to identify a metal's volatility temperature for a given set of incinerator conditions. The term 'volatility temperature' refers to the temperature at which the effective vapor pressure of a metal is 10^{-6} atm. The effective vapor pressure represents the combined vapor pressures of all species containing a metal. It reflects the quantity of the metal that would vaporize under a given set of conditions. A vapor pressure of 10^{-6} atm is selected because it represents a measurable amount of vaporization. The lower its volatility temperature, the more volatile a metal is expected to be. Volatility temperatures are a major parameter in a numerical partitioning model used to predict metal behavior in an incinerator.[27,28]

Average Trace Metal Discharge Distributions

Normalized distributions of each of the metals to the kiln ash (fraction of total discharge accounted for by kiln ash) are shown in Table 4 and are discussed below. Normalized fractions represent discharge distributions as they would have been had mass balance closure for the metals been 100%. Presentation of the data in this manner allows clearer data interpretation since mass balance closure is eliminated as a source of test-to-test data variability.[21]

Parametric Studies. In both the venturi/packed-column scrubber and single-stage ionizing wet scrubber tests, arsenic appeared to be much less volatile than expected; kiln ash accounted for greater than 80% of the discharged arsenic. Possible explanations for arsenic's behavior include the formation of a thermally-stable compound in the incineration environment or physical bonding of arsenic in the clay-based solid matrix.

Bismuth and cadmium were relatively volatile compared to the other trace metals. On average, less than 40% of the bismuth and cadmium were recovered in the kiln ash, compared to an average of greater than 75% of the arsenic, barium, chromium, copper, magnesium, and strontium.

[27] D. J. Fournier, Jr., L. R. Waterland, J. W. Lee, and G. J. Carroll, 'The Behavior of Trace Metals in Rotary Kiln Incineration: Results of Incineration Research Facility Studies', in 'Proceedings of the 17th Annual RREL Research Symposium, EPA/600/9–91/002, Apr. 1991.

[28] R. G. Barton, W. D. Clark, and W. R. Seeker, 'Fate of Metals in Combustion Systems', *Combust. Sci. Technol.*, 1990, **74**, 327.

Table 2 Nominal test conditions

	V/PC[a]	IWS[b]	Baird[c]	NBH[d]	CIC[e]	Drake[f]	SCP[g]
Kiln exit-gas temp./°C	825–928	819–929	832–994	824–985	982	546–829	818–987
/°F	(1517–1702)	(1507–1704)	(1541–1822)	(1516–1805)	(1800)	(1015–1524)	(1504–1808)
Kiln O_2 (%)	11.5	11.5	6.8–11.3	9–11.2	10	12.7–17	7.4–8
Afterburner temp. /°C	983–1196	1017–1163	1093	1208	1204	1093	1204
/°F	(1803–2184)	(1863–2125)	(2000)	(2206)	(2200)	(2000)	(2200)
After-burner O_2 (%)	7.5	7.5	7.5	6–7	8	8.7–11.8	7.5
Feed rate/kg h^{-1}	63	63	55	68	55	55	57
/lb hr^{-1}	(140)	(140)	(120)	(150)	(120)	(120)	(124)
Matrix	Spiked clay	Spiked clay	Soil	Sediment	Soil	Soil	Soil
Feed chlorine content (%)	0–8.3	0–6.9	<0.3	1	0.04	NA[h]	1.5
Kiln solids residence time (hr)	1	1	0.5	0.5	0.5	0.5	1

[a]Venturi/Packed-column Scrubber Tests.
[b]Single-stage Ionizing Wet Scrubber Tests.
[c]Baird–McGuire Tests.
[d]New Bedford Harbor Tests.
[e]Chemical Insecticide Corporation Tests.
[f]Drake Chemical Tests.
[g]Scientific Chemical Processing Tests.
[h]NA: Data not available.

Table 3 Feed metal concentrations (mg kg^{-1})

	V/PC^a	IWS^b	$Baird^c$	NBH^d	CIC^e	$Drake^f$	SCP^g
Arsenic	44	48	81–93	NA^h	794–1040	11–62	12–18
Barium	53	390	NA	NA	43–66	57–194	290–410
Bismuth	150	330	NA	NA	NA	NA	NA
Cadmium	8	10	NA	7	0.9–1.7	<1–2.0	29–97
Chromium	87	40	NA	161	13–19	12–20	190–270
Copper	470	380	NA	308	NA	35–49	6500–13 000
Lead	52	45	16–27	236	86–120	77–443	640–1220
Magnesium	17 200	18 800	NA	NA	NA	NA	NA
Strontium	280	410	NA	NA	NA	NA	NA

[a] Venturi/Packed-column Scrubber Tests.
[b] Single-stage Ionizing Wet Scrubber Tests.
[c] Baird–McGuire Tests.
[d] New Bedford Harbor Tests.
[e] Chemical Insecticide Corporation Tests.
[f] Drake Chemical Tests.
[g] Scientific Chemical Processing Tests.
[h] NA: Data not available.

Table 4 Range (average) of metal partitioning to kiln ash [% of discharged metal accounted for by ash]

	V/PC[a]	IWS[b]	Baird[c]	NBH[d]	CIC[e]	Drake[f]	SCP[g]	Overall[h]
Arsenic	84–94 (91)	80–95 (89)	36–76 (61)	NA[i]	NA	61–85 (73)	20–91 (53)	20–95 (74)
Barium	69–87 (77)	87–99 (95)	NA	NA	91–92 (92)	86–92 (89)	99	69–99 (90)
Bismuth	21–65 (32)	9–76 (41)	NA	NA	NA	NA	NA	9–76 (36)
Cadmium	<9–<30 (<15)	7–60 (25)	NA	8–61 (29)	22–76 (50)	72–78 (75)	3–61 (23)	3–78 (36)
Chromium	86–96 (93)	90–98 (94)	NA	88–92 (91)	68–91 (82)	72–80 (76)	91–99 (94)	68–99 (89)
Copper	58–98 (79)	77–95 (86)	NA	82–89 (85)	NA	58–64 (61)	66–93 (81)	58–98 (78)
Lead	6–84 (20)	71–92 (82)	69–93 (82)	19–53 (32)	88–93 (90)	68–90 (79)	9–56 (28)	6–93 (59)
Magnesium	99 (99)	99 (99)	NA	NA	NA	NA	NA	99
Strontium	82–94 (89)	94–99 (96)	NA	NA	NA	NA	NA	82–99 (94)
Mass Balance Closure	8–147 (71)	15–204 (70)	37–148 (84)	38–103 (71)	38–125 (69)	33–123 (79)	5–96 (40)	5–204 (69)

[a] Venturi/Packed-column Scrubber Tests.
[b] Single-stage Ionizing Wet Scrubber Tests.
[c] Baird–McGuire Tests.
[d] New Bedford Harbor Tests.
[e] Chemical Insecticide Corporation Tests.
[f] Drake Chemical Tests.
[g] Scientific Chemical Processing Tests.
[h] Range (average) across all tests.
[i] NA: Data not available.
[j] Percentage of feed metal accounted for by sum of kiln ash, scrubber liquor, and flue gas fractions.

Lead behavior was substantially different between the two tests. For the venturi/packed-column tests, the average recovery of lead in the kiln ash was 20%. For the single-stage ionizing wet scrubber tests, the fraction increased to 82%. Although the two tests were performed under the same nominal conditions, using essentially the same waste material, minor differences between the two programs may have combined with the sensitivity of lead to test variables (discussed below) to cause the wide variation in lead discharge distributions.

With the exception of arsenic and lead, metal behavior during the two parametric tests generally supported the use of volatility temperature in predicting relative partitioning.[27]

Superfund Treatability Tests. Partitioning behavior for barium and lead across the soil/sediment treatability tests was comparable to that during the parametric studies; barium was consistently non-volatile (with greater than 90% partitioning to the ash), while lead partitioning varied widely (from an average of 28% lead partitioning to the ash in the SCP tests to an average of 90% in the CIC tests).

Arsenic was more volatile in the treatability studies, although still not to the degree predicted; while individual tests had kiln ash arsenic fractions as low as 20%, on average greater than 50% of the discharged arsenic was found in the ash.

Cadmium behaved inconsistently across the tests. The New Bedford Harbor and SCP studies paralleled the parametric studies in that cadmium was relatively volatile; on average less than 30% of the discharge was represented by kiln ash. In contrast, cadmium was less volatile in the CIC and Drake tests; greater than 50% was discharged to kiln ash.

Chromium and copper exhibited some variability across the tests but remained predominantly non-volatile (greater than 60% discharge to the kiln ash), consistent with the parametric tests.

Effects of Incinerator Operating Conditions on Metal Distributions

Kiln Temperature. Kiln exit gas temperature was an operating variable in both of the parametric studies and, with the exception of the CIC tests, in each of the subject treatability studies. Over the range of temperatures tested there appeared to be no clear impact of kiln temperature changes on the partitioning of barium, chromium, copper, magnesium, or strontium. Increases in kiln temperature caused a relatively consistent increase in the volatilities of cadmium and lead (as evidenced by reduced kiln ash fractions of the metals). Increases in arsenic volatility were seen with increases in kiln temperature during two of the five tests for which data exist. An increase in bismuth volatility was evident with increases in kiln temperature during the single-stage ionizing wet scrubber tests.

An example of the impact of changes in kiln temperature may be seen in Figure 2. Cadmium, bismuth, and lead discharges are presented as a function of kiln temperature for the single-stage ionizing wet scrubber tests. As kiln temperature increases, a significant decrease in the kiln ash fractions of each of the three metals may be seen, with corresponding increases in the scrubber exit flue gas and scrubber liquor fractions. The impact of this variable on partitioning during the venturi/packed-column scrubber tests was less pronounced.[27]

106

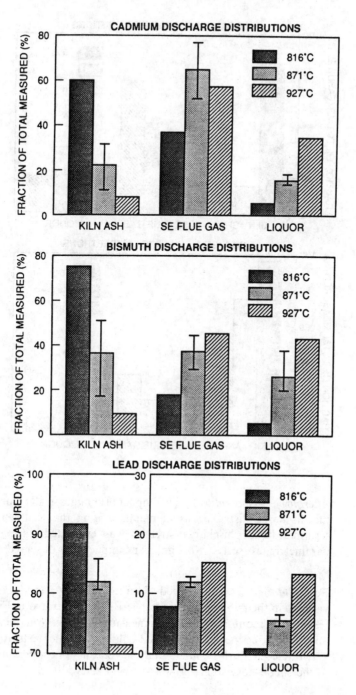

Figure 2 Effects of kiln temperature on the discharge distributions of cadmium, bismuth, and lead in the single-stage ionizing wet scrubber tests[27]

Figure 3 Effects of feed chlorine content on the discharge distributions of copper and lead in the venturi/packed-column scrubber tests[27]

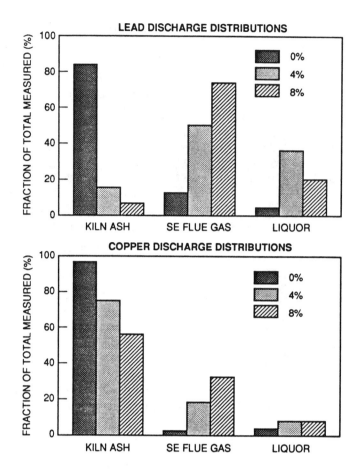

Afterburner Temperature. The impact of changes in afterburner temperature on the discharge distributions of metals among the scrubber exit flue gas and scrubber liquor discharge streams was evaluated during each of the two parametric test series. No significant impact was observed.

Feed Chlorine Content. Feed chlorine content was also evaluated as a test variable in the two parametric test series. As illustrated in Figure 3, increases in feed chlorine content resulted in a measurable increase in the volatilities of copper and lead during the venturi/packed-column scrubber tests. A similar relationship between feed chlorine content and the volatility of copper and lead was not clear in the single-stage ionizing wet scrubber tests.[27]

Lime Addition. The impact of blending lime with the waste feed was investigated during one of the four CIC treatability tests. Although the data from such limited testing should be interpreted with caution, results suggest that lime addition reduced the leachability of arsenic in both the untreated soil and in the kiln ash.

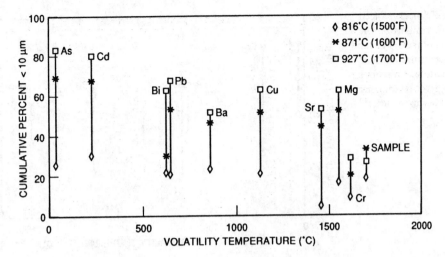

re 4 Effects of kiln
temperature on the
ons of metals in the
afterburner-exit flue
culate size fractions
single-stage ionizing
wet scrubber tests[27]

Additionally, lime addition appears to have reduced arsenic's volatility; kiln ash fractions of arsenic increased from 68–70% to 88% with lime addition.[24]

Particle Size Distributions[27]

The distribution of a given metal among flue gas particle size ranges is strongly influenced by the extent to which the metal vaporizes in the incineration system. Metals that do not vaporize significantly tend to be relatively evenly distributed in the flue gas particle size ranges on a per mass (mg kg^{-1} particulate) basis. Volatile metals tend to enrich in the fine particulate fractions, increasingly so with increased volatility.

The most complete assessment of particle size distributions, and of the sensitivity of the distributions to changes in operating conditions, took place during the single-stage ionizing wet scrubber tests. Figure 4 shows the fractions of the particulate metal found in the $<10\,\mu$m size range (as measured in the afterburner exit flue gas) during these tests. The fractions of the total particulate sample in this size range are also shown as are the effects of increased kiln exit temperature. Distributions are plotted against volatility temperatures to facilitate comparison of relative metal behavior.

With the exception of chromium, the average metal distributions in the flue gas particle size range of $<10\,\mu$m shifted from roughly 20% to an average of 60% as the kiln exit temperature was increased from 816 °C to 927 °C (1500 °F to 1700 °F). In addition, the redistribution of metals to this size range generally correlated with the predicted relative volatilities of the metals, with the volatile metals most affected.

In contrast to arsenic's distribution among the discharge streams (where it behaved as a relatively non-volatile metal and partitioned predominantly to the ash), arsenic in the flue gas behaved as the most volatile metal with respect to particle size redistribution; more than 80% of the arsenic particulate was found in the $<10\,\mu$m size fraction at a kiln temperature of 927 °C (1700 °F).

Figure 5 Effects of feed chlorine content on the distributions of metals in the afterburner-exit flue gas particulate size fractions in the single-stage ionizing wet scrubber tests[27]

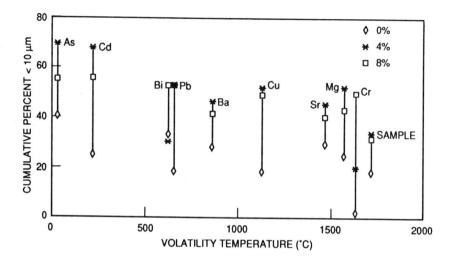

The effects of waste feed chlorine content on total-particulate and metal-specific size distributions are shown in Figure 5. When chlorine content was increased from 0% to 4%, the fraction of total particulate in the $< 10 \mu m$ fraction increased from 20% to approximately 35%. This is as expected if the presence of chlorine in the feed serves to increase the volatility of some feed inorganic constituents.

Chlorine had a pronounced effect on the particle size distributions of cadmium, chromium, copper, and lead, as reflected by a shift in particle size distributions of these metals greater than the shift observed for total particulate. For cadmium, copper, and lead, the shift to finer particulate occurred with the initial feed chlorine increase from 0% to 4%; the distribution to finer particulate increased from approximately 20% to approximately 50%. No additional redistribution occurred for these metals when the feed chlorine was further increased to 8%.

Chromium redistribution to finer particulate occurred with both increases in feed chlorine content; as chlorine increased from 0% to 4% to 8%, the fraction of chromium in the $< 10 \mu m$ fraction increased from 2% to 20% to 50%.

The impact of changes in feed chlorine content on the particle size distributions of copper and lead is consistent with the predicted increase in the volatility of the two metals in the presence of chlorine. Cadmium and chromium redistributions with changes in feed chlorine content were not similarly predicted.[27]

Mass Balance Closure

Mass balance closure is defined as the fraction of a feed metal recovered in the incinerator discharge streams (kiln ash, scrubber liquor, and scrubber exit flue gas). As indicated in Table 4, mass balance closure for individual metals ranged from 5% to 204% across the tests; test averages ranged from 40% to 84%, with an overall average of 69%. Though certainly not complete, it is consistent with past experience with combustion sources, in which metals mass balance closure has ranged from 30% to 200%.[29]

[29] US EPA, 'Trace Metals and Stationary Conventional Combustion Processes: Volume I—Technical Report', EPA/600/7-80/155a, 1980.

Possible explanations for incomplete closure include accumulation of metals in the incineration system (*e.g.* slag build-up in the afterburner; particulate accumulation in the ductwork; deposition of metals in the scrubber system) and incomplete liberation of metals in the digestion of solid samples.

7 Conclusions

Key conclusions based on the results of the subject tests include the following:

(i) Arsenic was much less volatile than predicted in both the parametric and Superfund treatability tests. Barium, chromium, and copper were also relatively non-volatile throughout the tests. The volatility of lead was inconsistent, as was that of cadmium to a lesser degree. In the parametric tests, bismuth was relatively volatile, while magnesium and strontium were non-volatile; these three metals were not evaluated in the treatability tests.

(ii) With the exception of arsenic and lead, relative volatilities were generally consistent with theoretical predictions.

(iii) In the majority of tests, increases in kiln exit gas temperature resulted in decreased kiln ash partitioning (increased volatility) of cadmium and lead. In approximately half of the tests for which data were available, increased kiln temperature also caused increases in the observed volatility of arsenic and bismuth.

(iv) Increases in afterburner temperature had no significant impact on the discharge distributions of metals among the scrubber exit flue gas and scrubber liquor discharge streams.

(v) Significant increases in the volatilities of copper and lead were observed with increases in feed chlorine content for one of two tests in which the variable was evaluated; results of the other test were inconclusive.

(vi) Based on the results of the single test in which it was evaluated, lime addition to the waste feed appears to have reduced the volatility of arsenic.

(vii) With the exception of chromium, the average metal distributions in the afterburner exit flue gas significantly shifted to finer particulate with increased kiln temperature; the degree of redistribution generally correlated with the predicted relative volatilities of the metals, with the volatile metals most affected.

(viii) Increases in feed chlorine content resulted in a redistribution of cadmium, chromium, copper, and lead to finer flue gas particulate.

What may appear to be conflicting results in the subject tests are more likely the result of the impact of a wide range of parameters on metal behavior. The IRF RKS studies have identified some variables which appear to influence partitioning. Additional variables are to be studied in upcoming bench-scale IRF research, the results of which may lead to further RKS evaluation.

Acknowledgements

Although the research described in this article has been funded wholly or in part by the United States Environmental Protection Agency through Contract No.

68-C9-0038 with Acurex Environmental Corporation, it has not been subjected to the Agency's review. Therefore it does not necessarily reflect the views of the Agency and no official endorsement should be inferred.

The substantial contributions of the management and staff of Acurex Environmental Corporation (operators of the IRF) to the research described in this article are appreciated. In particular, the author wishes to acknowledge the contributions of Dr Larry R. Waterland, Johannes W. Lee, and Donald J. Fournier, Jr. of Acurex to the completion of all phases of the tests, from planning to test execution to interpretation and reporting of results. The contributions of Robert C. Thurnau and Marta K. Richards of EPA/RREL to the IRF metal partitioning studies are appreciated as well.

The US Approach to Incinerator Regulation

E. M. STEVERSON

1 Introduction

Because of the recent research efforts of the Environmental Protection Agency
(EPA), more is known about United States (US) incineration technology and its
environmental performance than virtually any other waste management
alternative.[1] Research conducted by the EPA on incineration has shown that a
properly designed incinerator is a very effective treatment device that can be
operated safely and with negligible environmental impact or health risks.[2] The
research and regulatory framework provided by the EPA has also caused a
significant improvement in the practice of incineration over the past decade.
These improvements include better performance and control, reduced emissions,
and standardized sampling and analytical methods.

In the US, industry and government have accepted incineration as the
treatment method of choice for many types of waste. The many benefits of
incineration include the destruction of toxic organics, volume reduction,
potential energy recovery, wide range of application, and homogenization of the
waste stream. Despite a generally negative public attitude toward incineration,
the US EPA has advocated incineration as the preferred treatment method for a
wide range of waste streams. The EPA has endorsed incineration because it has in
place, a regulatory framework that effectively controls the emissions resulting
from the operation of incinerators. The following chapter describes the
developmental history of US incinerator regulations, the current regulatory
approach and basis, the major laws and regulations, their impact on the industry,
and the future direction of the regulations.

History of US Incinerator Regulations

The history of US incinerator regulations and of environmental regulations in
general, is fairly short. The first type of incineration widely used in the US was for
the burning of common refuse or Municipal Solid Waste (MSW). In 1921, there
were more than 200 MSW incinerators in the US, but the first federal standards

---

[1] American Society of Mechanical Engineers, 'Hazardous Waste Incineration: A Resource
Document', ASME Publications, New York, 1988.
[2] C.R. Dempsey and E.T. Oppelt, *J. Air Waste Manage. Assoc.*, 1993, **43**, 25.

controlling MSW incinerator emissions were not enacted until 1970.[2] Even then, only a particulate emission limit was established and substantive standards governing MSW incineration were not enacted until 1991. Since the practice of MSW incineration has for most of its history been unregulated, the degree of emissions control has in the past been inadequate. The current perception of incineration has apparently been influenced by past practices because today's incineration industry is plagued by a lack of public acceptance.

The history of federal air pollution statutes dates back to 1955 with the passage of the Air Pollution Control Act. This law provided funding for federal research and technical assistance to the states and local governments for air pollution control. Similar air pollution control legislation known as the Clean Air Act (CAA) was passed in 1963, and amended in 1965, 1966, 1967, and 1969. In this same time period, public awareness of the need for pollution control was influenced by the book 'Silent Spring', by Rachael Carson. This lead to three important developments in 1970: the establishment of the EPA, the passage of National Environmental Policy Act (NEPA), and a major amendment of the CAA.[3]

NEPA set forth the national environmental policy and direction with the following stated purpose:

> '. . . encourage productive and enjoyable harmony between man and his environment; to promote efforts which will prevent or eliminate damage to the environment and biosphere and stimulate the health and welfare of man; to enrich the understanding of the ecological systems and natural resources important to the Nation; and to establish a Council on Environmental Quality.'

NEPA called for a predecisional assessment of the environmental impact of all major federal actions in an Environmental Impact Statement (EIS). While NEPA only required an EIS for federal activities, many states adapted the EIS requirement for all major activities. NEPA also became a template for subsequent US pollution control legislation.

In December 1970, the EPA was established to develop regulations consistent with the environmental statutes enacted by Congress. One of the first sets of regulations developed by the EPA was in response to the CAA Amendments of 1970. These amendments established National Ambient Air Quality Standards (NAAQS) that set permissible levels for the following pollutants: particulate, sulfur dioxide (SO_2), carbon monoxide (CO), nitrogen dioxide (NO_2), and ozone (lead was added in 1987). The permissible levels for ambient air were to be attained within five years (by July 1975) by enforcement through implementation plans established by the states.

More realistic NAAQS attainment deadlines were established by further amendments to the CAA in 1977.[4] These amendments also established maximum emission standards for new stationary sources and major modifications to existing stationary sources. These standards, called New Source Performance Standards (NSPS), were applied on an industry specific basis, which included the MSW incineration industry. The NSPS established a particulate emission limit that represented the first federal standards directly controlling incinerator emissions.

---

[3] W. S. Rickman, 'Handbook of Incineration of Hazardous Waste', CRC Press, Boca Raton, 1991.
[4] C. A. Wentz, 'Hazardous Waste Management', McGraw Hill, New York, 1989.

The 1977 amendments also established National Emission Standards for Hazardous Air Pollutants (NESHAPS). These provided control for airborne hazardous chemicals that are known to present a public health risk at relatively low concentrations. The original standards included asbestos, benzene, beryllium, mercury, and vinyl chloride. Arsenic, sulfuric acid, and radioactive isotopes were added latter. Any stationary source, including incinerators that emitted these pollutants, was subject to these emission standards.

In 1990, the CAA was amended in what has been widely acclaimed as the most comprehensive and sweeping environmental law ever enacted. Whether this is true or not, the 1900 amendments did completely overhaul the act and resulted in a more detailed regulatory framework for incinerators. These standards represent an important part of the current US regulatory approach and will be covered in detail later in this chapter.

The most comprehensive set of incinerator regulations in the US is not the result of air pollution control laws but of solid waste laws. The first federal statute covering the management of solid waste was the Solid Waste Disposal Act of 1965. This act financed a research program aimed at improving the collection, transport, recycling, and disposal of solid waste. Increased public awareness of the need to control pollution lead to the passage of the Resource Recovery Act in 1970, which expanded the federal role.[5] Then, the discovery and widespread publicity of major environmental catastrophes resulting from improper past disposal practices heightened public awareness of hazardous waste. This awareness lead to activism that resulted in major legislation to regulate waste management, known as the Resource Conservation and Recovery Act (RCRA) of 1977.

While MSW incinerators have been operated for at least 100 years in the US, hazardous waste incinerators have only been in existence for about 20 years. The comprehensive regulations resulting from RCRA established, for the first time, technical standards for hazardous waste incinerators. These standards include both performance and operating requirements and established a process through which all hazardous waste incinerators must obtain a permit. These standards represent an important part of the current US regulatory approach and will be covered in detail later in this chapter.

In 1976, the Toxic Substances Control Act (TSCA) was enacted to regulate toxic chemical substances. The management of toxic substances after they become waste is generally regulated by RCRA. However, the management of polychlorinated biphenyls (PCB) is regulated by TSCA. TSCA mandates that certain forms of PCB contaminated material must be incinerated and establishes a permitting program for PCB incinerators.

The Current Regulatory Approach

The major regulatory programs applicable to the various types of US incinerators include those that regulate air and water pollution, solid and hazardous waste management, toxic substances, and remedial action. Many regulatory programs

---

[5] L. Theodore and J. Reynolds, 'Introduction to Hazardous Waste Incineration', John Wiley and Sons, New York, 1987.

Table 1 Potentially applicable laws

Incinerator type	Potentially applicable laws and regulations
Municipal solid waste	CAA, CWA
Medical waste	CAA, CWA
Sewage sludge	CAA, CWA
PCB waste	CAA, CWA, TSCA
Remedial action (soil decontamination)	CAA, CWA, TSCA, RCRA, CERCLA
Hazardous waste	CAA, CWA, TSCA, RCRA

CAA—Clean Air Act
CERCLA—Comprehensive Environmental Response, Compensation, and Liability Act
CWA—Clean Water Act
RCRA—Resource Conservation and Recovery Act
TSCA—Toxic Substances Control Act

[e.g. CAA, Clean Water Act (CWA), RCRA] are established at the federal level but administered and enforced at the state level. The programs implemented at the state level have to be at least as restrictive as the federal laws, but can be more restrictive. Some programs (e.g. CAA) may be further delegated by the states to local governments, who may also invoke more restrictive controls. Other programs (e.g. TSCA) are administered and enforced at the federal level. As a result, an incinerator operator in the extreme case, has to deal with as many as three levels of government and five regulatory programs.

The applicable major US laws that regulate various types of incineration are shown in Table 1. Because so many different laws are potentially applicable to an incinerator application and because these laws are very complex, establishing and maintaining compliance requires a large investment of time, money, and effort. In most instances, several permits issued by independent agencies are required. If the processes required to obtain the various permits are not carefully coordinated, permits with conflicting operating conditions may result. Depending on its location, an incinerator may also be subject to local standards that may be more restrictive than federal and state standards.

Many incinerators employ wet air pollution control systems (i.e. control equipment such as venturis that use water) that discharge a wastewater stream. In order to legally discharge these streams, wastewater discharge permits under the CWA are required. Another law that applies when an incinerator is used to clean up a contaminated soil site, is the Comprehensive Environmental Response, Compensation, and Liability Act (CERCLA). This law, administered at the federal level, sets forth the standards governing the clean-up of hazardous waste contaminated land areas.

The three major US statutes that are directly applicable to incinerators are the CAA, RCRA, and TSCA. These programs will be described in detail in the following section. By order of these statutes, the EPA has developed performance standards, operating requirements, and standardized emissions measurement procedures that represent the controls instituted by the US Government to protect the public health and environment from incinerator emissions. The performance standards discussed in this chapter are either technology-based, health-based, or risk-based. For the purposes of this discussion, these terms are defined as follows:

> *Technology-based*: based on the potential achievement of the state-of-the-art in technology;
> *Health-based*: based on requirements to protect the public from deleterious health effects (usually applies to non-carcinogens);
> *Risk-based*: based on requirements to protect the public from an additional defined risk level (*e.g.* 1 in 100 000) of death due to cancer (usually applied to carcinogens).

2 Major US Statutes and Regulations

The Clean Air Act Amendments of 1990

The CAA and its amendments are a very complex and multifaceted law. The CAA has three major sets of provisions applicable to incinerators known as NAAQS, NSPS, and NESHAPS. Some of these provisions have standards that are defined specifically for certain types of incineration and others apply to all sources. The CAA also establishes a state permitting program for stationary sources of air pollution. The permitting process occurs in two steps, a construction permit and an operating permit. The construction permit is issued prior to construction. The operating permit is issued after compliance with the applicable emissions standards is demonstrated in a compliance test.

National Ambient Air Quality Standards. The NAAQS established ambient air quality standards for the following pollutants: CO, SO_2, nitrogen dioxide, particulate matter, ozone, and lead. To enforce the standards, the entire country was divided into Air Quality Control Regions and the air quality in each region is monitored. Based on the monitoring, the regions are either designated as attainment or non-attainment areas for each pollutant. The emission standards that apply to an incinerator are greatly affected by the attainment status of the chosen site.

If a *new* incinerator that is considered to be a 'major emitting facility' for a regulated pollutant is to be located in an *attainment* area for the pollutant, the permitting process must include a Prevention of Significant Deterioration (PSD) Review. The PSD Review is an assessment of the plant's potential to deteriorate the air quality of the area. An incinerator is considered a major emitting facility if it is a MSW incinerator capable of charging greater than 225 tonne d^{-1} and it has the potential to emit 90 tonne y^{-1} or more of a regulated pollutant. Any type of incinerator is also a major emitting facility if it has the potential for emitting 225 tonne y^{-1} or more of a regulated pollutant.

A PSD Review is also required if an *existing* incinerator in an *attainment* area is to be modified and the modification is considered to be significant in its potential to degrade the air quality of the area. The regulations establish emission thresholds for each regulated pollutant, which determine whether the modification is significant.

If a PSD Review is required, a construction permit will not be granted until the applicant demonstrates that the 'Best Available Control Technology' for the pollutant is implemented. Best Available Control Technology is defined as '. . . an emission limitation based on the maximum degree of reduction of each pollutant . . . which the permitting authority . . . determines is achievable.' In addition, an assessment must be made of the following:

   (i) the ambient air quality in the vicinity of the source;
  (ii) the potential impact on soils and vegetation in the area;
 (iii) the air quality impacts that could be realized due to growth caused by the new source or modification; and
 (iv) the visibility impacts of the project.

The applicant must also demonstrate that it will not cause or contribute to a violation of the NAAQS or any allowable incremental increase in the ambient concentration level of a pollutant established for the area.

Trying to locate a *new* major emitting incineration facility or significantly modifying an *existing* incineration facility in or near a *non-attainment* area can be very difficult and will likely not be cost effective. In order to get a construction permit, the applicant must show that the project will improve air quality in the area. To obtain a permit, the following must be demonstrated:

   (i) existing sources in the area will reduce their emissions by an amount greater than that emitted by the proposed project;
  (ii) the Lowest Achievable Emission Rate is accomplished without regard to cost; and
 (iii) all of the owner's other sources in the state are in compliance with their permits.

These requirements ensure against any further deterioration of air quality in a non-attainment area.[6]

New Source Performance Standards.   The NSPS establish emissions standards for certain pollutants from MSW incinerators and any type of incinerator that charges more than 45 tonne d^{-1} of waste. Standards under NSPS are developed to reflect the maximum achievable reduction of air pollutant emission with consideration to cost, any non-air quality health and environmental impacts, and energy requirements. The NSPS standards for various solid waste incinerators are summarized in Table 2. The application of the limits to MSW incinerators depends on the size and type of unit as indicated in the table.[7]

National Emission Standards for Hazardous Air Pollutants. Under the CAA of 1970, EPA regulated the emissions of eight air toxics under NESHAPS.[8] In the CAA amendments of 1990, Congress revised NESHAPS and designated 189 air

[6] US Code of Federal Regulations, Title 40, Part 50.
[7] US Code of Federal Regulations, Title 40, Part 60.
[8] US Code of Federal Regulations, Title 40, Parts 61 and 63.

Table 2 New source performance standards for incinerators[a]

Capacity, tonne d^{-1}	New source performance standards	Emission guidelines for existing facilities	
	Unit >225	Unit >225 ≤ 1000	Facility >1000
Particulate matter [mg (Nm^3)$^{-1}$]	34	69	34
Opacity, %	10	10	10
Total chlorinated PCDD plus PCDF [ng (Nm^3)$^{-1}$]			
—Mass burn units	30	125	60
—RDF fired units	30	250	60
Acid gas control % reduction or emissions (p.p.m.)			
HCl	95 (25)	50 (25)	90 (25)
SO$_2$	80 (30)	50 (30)	70 (30)
NO$_x$	(180)	None	None
CO, p.p.m.	50–150[b]	50–250[b]	50–250[b]

[a]All emissions limits are referenced to dry gas conditions at 7% oxygen concentration.
[b]Range of values reflect differing types of MSW incinerators.
RDF—Refuse Derived Fuel
PCDD— Polychlorinated Dibenzo-*p*-dioxin
PCDF—Polychlorinated Dibenzofuran

toxics to be regulated under what is now called the Hazardous Air Pollutant (HAP) Program. The list includes air toxics in the form of organic chemicals, toxic metals, and radionuclides. Within eight years, the EPA is directed to establish HAP emission standards for more than 200 industry categories.

The new emission standards to be established are to be technology-based and if deemed necessary for a particular pollutant, health-based. The standards will be applied differently depending on whether the source is new or existing or if it is considered a major or area source. A 'major source' is one that emits or has the potential to emit considering air pollution controls, 9 t y^{-1} or more of any HAP or 23 t y^{-1} or more of any combination of HAPs. An 'area source' is a HAP source that is not a major source.

Applicable to the major sources are Maximum Achievable Control Technology (MACT) standards. MACT is defined as the maximum achievable reduction of air pollutant emission with consideration to cost, any non-air quality health and environmental impacts, and energy requirements. These standards may include the required application of operational standards (including operator training certification), processes, treatments, or practices, which reduce or eliminate the volume of emissions. The most stringent MACT standards will be applied to *new* major sources. For new major sources, the degree of emission reduction shall not be less stringent than the emission control that is achieved in practice by the best controlled similar source as determined by the permitting authority. Somewhat less stringent standards will be applied to existing sources, but the standards must be representative of the best similar controlled sources as determined by the permitting authority.

Applicable to area sources, which may include some types of incineration, will be Generally Achievable Control Technology (GACT) standards. Because area sources may include small businesses, such as gas stations and dry cleaners, the EPA may consider economic impacts and the technical capabilities of the category when developing GACT standards. In addition to the means of reducing emissions defined under MACT, GACT may include management practices.

Since MACT standards are technology-based, the emission of a HAP after the application of a MACT standard may not be protective of human health. In order to protect the public against this possibility, the 1990 amendments included a residual risk provision. Under this provision, the EPA is directed to assess the residual risk remaining after the MACT standards are implemented. The measured residual risk will be presented to Congress by 1996 in a report along with other factors including:

(i) the methods of calculating residual risk;
(ii) the health significance of the residual risk;
(iii) commercially available methods and cost of reducing risk;
(iv) the implications to actual health effects of persons in the vicinity of sources; and
(v) recommendations on legislation regarding the residual risk.

If Congress does not act on recommendations within eight years of the report, the EPA is directed to promulgate more stringent standards to protect the public. As a result, incinerator operators that implement MACT may soon thereafter, have to implement more stringent controls.

Resource Conservation and Recovery Act

General Provisions. RCRA set forth the program by which hazardous waste became regulated in the US. The regulations resulting from RCRA represent what is known as a 'cradle to grave' approach to hazardous waste management. Under this approach, the waste generators, transporters, and the treatment, storage, and disposal facilities are all regulated.

Hazardous waste incinerators are regulated as treatment facilities under RCRA. RCRA regulates much more than the operation and emissions of a hazardous waste incinerator. The permit for an incinerator under RCRA covers almost every aspect of the facility including waste acceptance, handling, and residue disposal.

The objective of the act and regulations is to ensure that hazardous waste is managed in a manner that protects human health and the environment. The major provisions of RCRA include the following:

(i) identification and listing of hazardous waste;
(ii) standards applicable to generators;
(iii) standards applicable to transporters;

120

    (iv) standards applicable to owners and operators of treatment, storage, and disposal facilities;

    (v) standards applicable to specific hazardous waste and specific types of facilities;

    (vi) land disposal restrictions; and

    (vii) permitting program.

The passage of RCRA was instigated by the discovery and publicizing of abandoned hazardous waste sites across the country. The government became a financial party to the cleanup of many of these sites and it was clear that under the current law, land disposal was the preferred disposal method for hazardous waste and no incentive was given to reduce or treat waste. Of course, the unregulated disposal of hazardous waste on land is a clear threat to human health by providing many pathways of exposure including contamination of drinking water supplies. As stated in the act, requiring that hazardous waste be properly managed in the first instance reduces the need for corrective action at a future date.

RCRA establishes a hierarchy of priorities in hazardous waste management. At the top of this priority list is waste elimination at the source. But since this is not always possible, the next priorities deal with waste management and include in order, waste reduction, recycling, reuse and recovery, treatment, and residual disposal. These priorities were substantiated in 1984 through a major amendment to RCRA.

RCRA provided that any state may administer and enforce hazardous waste regulations and permits. The states develop a program and apply for authorization to the EPA. The state's program must meet the minimum requirements of the federal program, but may be more stringent. If the state's program is approved, permitting and regulatory authority is transferred to the state. Until a state receives authority, the federal EPA will apply the federal program in that state.

Land Disposal Restrictions. The 1984 amendments, known as the Hazardous and Solid Waste Amendments, set forth a schedule for the complete prohibition of land disposal for untreated hazardous waste. Now, except for some specific waste streams that have been granted extensions, no hazardous waste may be land disposed until it has been treated to minimum treatment standards.

The treatment standards are specific to various broad groupings of waste types. The treatment standards are established as either a specific technology (or group of technologies) or as a performance level (*i.e.* the concentration level of a hazardous constituent in the waste or an extract from the waste). These treatment standards were important to the US hazardous waste incineration industry because incineration was designated as the required treatment technology for a large number of wastes.[9]

Hazardous Waste Identification. As defined by RCRA, a waste is hazardous if it exhibits specific hazardous characteristics or if it is found in one of several lists of wastes from specific and non-specific sources. The hazardous characteristics that define 'characteristic wastes' include ignitability, corrosivity, reactivity, and

---

[9] US Code of Federal Regulations, Title 40, Part 268.

toxicity, which are determined by specific tests. The waste lists that define 'listed wastes' include wastes from specific sources, non-specific sources, discarded and off-specification chemical products, and container residues.

An important provision of the identification of hazardous waste to incinerator operators is that residues resulting from the treatment of hazardous waste are also a hazardous waste. This means that ash and wastewater residues from the incineration of a listed waste are also hazardous wastes. The residues from the incineration of characteristic wastes are hazardous waste until it is proven that the waste no longer exhibits a hazardous characteristic.[10]

Treatment, Storage, and Disposal (TSD) Facility Standards. The standards for owners and operators of hazardous waste TSD facilities set forth financial assurance, safety, record-keeping, and operating requirements. In addition, specific operating practices and performance standards are defined for specific waste management practices such as container and tank storage, land disposal, and incineration. A hazardous waste incineration facility will in most cases include facilities for container and tank storage and other treatment processes such as ash stabilization. Often the complex will also include a land disposal facility for treated residues. As a result, many hazardous waste incineration facilities will have to comply with most of the provisions of this set of standards.

The facility standards are very comprehensive. All TSD facilities are required to have numerous plans and procedures in place that are designed to ensure that the facility is properly staffed, trained, and equipped to safely manage hazardous waste. The facility must have adequate fire suppression, spill control, and communication equipment. It must also have a training program to ensure that the staff are capable of safely and responsibly operating the facility. An inspection program must be implemented to ensure that all equipment is properly functioning, all safety supplies are available, and that all of the waste is contained. A facility operating record must track all waste into and out of the facility and demonstrate that all of the programs, such as inspections, are being performed regularly.

The standards also require that the facility has plans in place that require the proper characterization of waste. The characterization requirements must include all the information necessary to allow the safe handling, storage, and treatment of the waste. The facility must also have a contingency plan that clearly outlines the steps taken in response to spills, releases, or other emergencies. The steps must include notification of state and local authorities.

The facility must also have in place a plan for the closure of the facility at the end of its operating life. This plan includes the steps taken to decontaminate, dismantle, dispose of debris and remaining waste, and verify the cleanliness of the closed site. The facility must also have financial mechanisms in place to assure that adequate funds are available to properly close the site.[11]

Permitting Program. The backbone of RCRA is its permitting program. Under RCRA, all new hazardous waste TSD facilities must obtain a permit prior to

[10] US Code of Federal Regulations, Title 40, Part 261.
[11] US Code of Federal Regulations, Title 40, Part 264.

construction and operate in compliance with the terms of their permit. The RCRA incinerator permitting process for new incinerators is a highly complex, costly, and lengthy process that takes place in four phases. One aspect of the process that often adds considerable complexity is public involvement, which is discussed separately.

Hazardous waste incinerators that were *in existence* when the regulations were promulgated were allowed to continue operating under 'interim status' until they received a permit. Under interim status, the owners or operators submitted an application, upgraded the facility as necessary to meet the standards, and performed a compliance test known as a 'trial burn'. If the trial burn demonstrated compliance, a permit was issued. Under the law, the EPA was required to issue or deny a final permit to all existing hazardous waste incinerators by November 1989.

The first step in the permitting process for a *new* incinerator is the submission of a permit application by the owner of the incinerator. The application must provide very detailed design information and a plan for a trial burn. After review and comment by the permitting authority, a permit for the incinerator construction is issued.

The initial or shake-down phase of operation under a permit begins immediately after construction. In this first phase, the operator is allowed 720 h of operations on hazardous waste to identify mechanical difficulties and ensure that the unit is ready for steady-state operation. The burning of fuel or non-hazardous waste is not restricted during this period.

After the shake-down phase is successfully completed, the second or trial burn phase begins. This phase allows only for the incinerator compliance test to be completed. The trial burn test is used to show compliance with the incinerator performance standards. Within 90 days of completion of the trial burn, the results of the test must be submitted to the permitting authority.

The third phase begins after the trial burn is complete and runs until the permitting authority issues the 'finally effective RCRA permit'. During the third phase, the facility may operate on hazardous waste at specified conditions.

The permitting authority reviews the results of the trial burn to determine if the incinerator is capable of complying with the performance standards. If compliance was not shown, the permit must be modified to allow for another trial burn. If compliance was shown, the permit will be modified to set the final operating requirements that are consistent with the trial burn conditions. The finally effective RCRA permit is then issued with a life of no longer than 10 years.

The entire incinerator permitting process from submission of the initial application until a final operating permit is obtained normally takes at least three years. In some cases, particularly where public concern for the facility is substantial, the process has taken seven years or more. The cost of the process is also an important consideration. The permitting process for a large commercial facility, including the trial burn, will normally require an investment ranging from five to ten million US dollars.

The permit application for a hazardous waste incinerator is a very detailed document. The document will typically consist of five to ten loose-leaf binders (8 cm) of information including text and detailed engineering drawings. The document will extensively cover the following general subjects: site characteristics; anticipated or known waste characteristics; waste analysis plans; detailed engineering descriptions; trial burn plan; emergency preparedness and prevention, fugitive emissions control; inspection and maintenance; contingency plans, operating training; facility closure; and groundwater monitoring.

The trial burn plan will present the waste or surrogate waste feed properties, operating conditions, sampling and analysis requirements, and quality assurance/quality control procedures for the test.[12]

Performance Standards and Operating Requirements. The trial burn is used to demonstrate compliance with the RCRA incinerator performance standards. The trial burn is conducted under highly controlled test conditions on well characterized waste. During the trial burn, the operating conditions are closely controlled, continuously monitored, and recorded. The trial burn also includes extensive stack sampling and continuous emissions monitoring of off-gas constituents including CO and oxygen (O_2).

The performance standards and operating requirements for hazardous waste incinerators are published in the federal regulations.[11] However, what is currently published in the regulations is a very small part of what is actually demanded by the regulators. The existing RCRA hazardous waste incinerator regulations include the following technology-based performance standards:

(i)   99.99% Destruction and Removal Efficiency (DRE) for each Principal Organic Hazardous Constituent (POHC) in the waste feed as calculated by the following equation:

$$DRE = \frac{W_{in} - W_{out}}{W_{in}} \times 100 \qquad (1)$$

where W_{in} is mass of POHC fed to the incinerator and W_{out} is mass of POHC emitted at the stack;

(ii) 99.9999% DRE for dioxins and furans (or POHCs more difficult to destroy than dioxins and furans), if waste containing these substances will be burned;

(iii) 99% removal of hydrochloric acid (HCl) from the stack gas if emissions are greater than $1.8\,kg\,h^{-1}$; and

(iv) particulate emissions must be controlled to less than $180\,mg\,(Nm^3)^{-1}$ corrected to 7% O_2.

The regulations also specify operating requirements when the incinerator is burning hazardous waste. These include the following: systems to control fugitive emissions from the combustion chamber (*e.g.* seals or maintaining less than atmospheric pressure); continuous monitoring of temperature, waste feed rate, and an indicator of incinerator residence time; continuous monitoring of carbon monoxide; and a functioning system to automatically cut off waste feed to the incinerator when operating conditions deviate from limits established in the permit.

The regulations also specify that acceptable operating limits will be established

---

[12] US Code of Federal Regulations, Title 40, Part 270.

124

in the permit for carbon monoxide, waste feed rate, combustion temperature, and residence time. These will be established from the trial burn results.

Destruction and Removal Efficiency. The DRE standard is the backbone of the RCRA incinerator standards. The EPA, based on the waste characteristics and recommendations by the incinerator operator, specifies one or more POHCs for the trial burn. POHCs are approximately 400 recognized hazardous chemical constituents found in the RCRA regulations.[10] The POHCs have been ranked in order of their theoretical thermal stability. Two ranking schemes are currently used by the EPA and usually one thermally stable POHC is chosen from each list. One list is simply based on the theory that the lower the heat of combustion, the more difficult an organic compound is to destroy. The other list was developed by experimentation, which determined the temperature necessary to destroy 99% of each compound in two seconds residence time.

The RCRA regulatory approach is based on the premise that, under similar operating conditions, if the destruction of a thermally stable compound is demonstrated, then all less stable POHCs will also be destroyed. After a successful trial burn, the permittee will be allowed to burn wastes contaminated with the trial burn POHCs and all less stable POHCs. As a result of these rankings, incinerator owners wanting a great deal of operational flexibility typically designate carbon tetrachloride and benzene as their POHCs. These compounds are ranked fourth and third on the heat of combustion and temperature indices, respectively.[13]

During a trial burn, the incinerator is operated under 'worst case' conditions of minimum combustion temperature, minimum residence time, maximum chamber pressure, and maximum waste feed rate. These conditions become the limits during all subsequent operations after a successful trial burn is completed. Similar operating conditions are established for air pollution control equipment parameters based on the trial burn. These may include for example, minimum venturi pressure drop, minimum liquid-to-gas flow ratios, and maximum scrubber liquid pH.

The RCRA amendments of 1984, called for the EPA to establish regulations for the burning of hazardous waste in Boilers and Industrial Furnaces (BIFs). Industrial furnaces include devices such as cement, lime, and aggregate kilns. The BIF regulations became effective in August 1991 and closed a loophole that had allowed the unregulated burning of hazardous waste in these devices. The promulgated BIF regulations are presently more stringent than the existing hazardous waste incinerator regulations. They include risk-based standards for toxic metals and required controls on particulate, organics, chlorine (Cl_2), and HCl.[14]

When the EPA developed the BIF regulations, they were also in the process of revising the hazardous waste incinerator regulations. The EPA believes that regulations for incinerators and BIFs should be technically consistent, if not entirely the same.[15] While new regulations have been promulgated for BIFs, they

---

[13] US Environmental Protection Agency, 'Technical Implementation Document for EPA's Boiler and Industrial Furnace Regulations', EPA-530-R-92-011, 1992.

[14] US Code of Federal Regulations, Title 40, Part 266.

[15] US Environmental Protection Agency, *Fed. Reg.*, 1990, **55**, 17 869.

have not yet been for incinerators. RCRA, however, allows the EPA to impose any permit condition deemed necessary to protect human health and the environment (known as the 'omnibus authority'). The BIF regulations, therefore, are presently used by the EPA and states to regulate hazardous waste incinerators.

The old (incinerator) and new (BIF) regulatory schemes are given in Table 3 for comparison. The primary difference is the addition of standards to control Products of Incomplete Combustion (PICs), Cl_2, and toxic metals. The new scheme also makes use of risk-based standards in addition to technology-based standards.

Products of Incomplete Combustion.   The old regulatory scheme addressed the destruction of organics by the DRE standard, which has some inherent limitations. The standard does not control the total mass of POHC emitted since 0.01% of all POHCs fed are allowed to be emitted regardless of the feed rate. In addition, the standard does not directly control the emission of PICs, which may be as or more toxic than the original organic species burned.

The EPA determined that the most effective way to limit PIC emissions was by making sure the incinerator was operating at a high combustion efficiency. The EPA considers CO to be the best available indicator of combustion efficiency and a conservative indicator of upset conditions. The EPA has data that indicate that PIC emissions do not pose a significant health risk when an incinerator is operated such that CO emissions are less than 100 p.p.m. (corrected to 7% O_2, dry).[16] Incinerators, therefore, must continuously monitor CO and O_2 and maintain the corrected CO at less than 100 p.p.m. on a one-hour rolling average basis.

The EPA also believes that PIC emissions may be acceptable when CO emissions are greater than 100 p.p.m. So, it provided two alternative standards for those facilities that cannot maintain CO below 100 p.p.m. By one standard, the operator must continuously monitor total hydrocarbon emissions and the corrected emissions must be less than 20 p.p.m. If the operator cannot meet this standard, compliance can be shown by measuring organic emissions during a trial burn. If the maximum ground-level concentrations (as determined by dispersion modeling) of specific organics do not exceed defined risk-based limits, the facility is in compliance. The corrected CO and hydrocarbon emissions measured in the trial burn then become the alternative limits.

Toxic Metals and Chlorine.   Under the old regulatory scheme, toxic metals were assumed to be controlled by the particulate standard. After evaluating the potential risk of metals, the EPA determined that additional controls were necessary. The BIF regulations include risk-based emission limits for the twelve toxic metals shown in Table 3. An incinerator operator may choose to comply with the standards by one of three methods:

(1) compliance with the conservative limits given in Table 3 as metal *feed* limits (allow no credit for removal by APCE);

[16] US Environmental Protection Agency, *Fed. Reg.*, 1990, **55**, 17 882.

126

Hazardous waste
~~incin~~erator regulations

Emission constituent	Incinerator limits	Boiler and industrial furnace limits
Particulate $[\text{mg}\,(\text{Nm}^3)^{-1}]^a$	180	180
CO^a	Continuous monitoring	Continuous monitoring and <100 p.p.m. or; >100 p.p.m. and <20 p.p.m. hydrocarbons or; >100 p.p.m. and acceptable risk
Hydrocarbon	NR	Continuous monitoring if $CO > 100$ p.p.m.
Dioxin and furan	NR	Incinerators with dry particulate control devices operated at 230–$400\,°C$ must show acceptable risk
Antimony[b,c]	NR	14–$31\,000\,\text{g}\,\text{h}^{-1}$
Barium[b,c]	NR	2400–$5\,000\,000\,\text{g}\,\text{h}^{-1}$
Lead[b,c]	NR	4.3–$9200\,\text{g}\,\text{h}^{-1}$
Mercury[b,c]	NR	14–$31\,000\,\text{g}\,\text{h}^{-1}$
Nickel[b,c]	NR	950–$2\,000\,000\,\text{g}\,\text{h}^{-1}$
Selenium	NR	190–$410\,000\,\text{g}\,\text{h}^{-1}$
Silver[b,c]	NR	140–$310\,000\,\text{g}\,\text{h}^{-1}$
Thallium[b,c]	NR	14–$31\,000\,\text{g}\,\text{h}^{-1}$
HCl[b,c]	$1.8\,\text{kg}\,\text{h}^{-1}$ or 99% removal	330–$720\,000\,\text{g}\,\text{h}^{-1}$
Cl_2[b,c]	NR	19–$41\,000\,\text{g}\,\text{h}^{-1}$
Arsenic[b,d]	NR	0.11–$240\,\text{g}\,\text{h}^{-1}$
Beryllium[b,d]	NR	0.26–$580\,\text{g}\,\text{h}^{-1}$
Cadmium[b,d]	NR	0.04–$86\,\text{g}\,\text{h}^{-1}$
Chromium[b,d]	NR	0.2–$430\,\text{g}\,\text{h}^{-1}$

[a] Corrected to 7% O_2.

[b] The ranges given are the lowest and highest levels, which depend upon local land use, stack height, and terrain. Less stringent limits may be established by site-specific dispersion modeling.

[c] The feed rate or emission rate must be maintained below the screening limits.

[d] The sum of the ratios of the actual feed or emission rates to the screening limits for each metal must be maintained to less than or equal to 1.0.

NR—Not regulated.

(2) compliance with the conservative limits given in Table 3 as metal *emission* limits (allow credit for APCE removal);

(3) compliance with metal feed or emissions limits established through site-specific air dispersion modeling and risk assessment (may be established with or without credit for APCE removal).

The only difference between methods one and two is that for compliance with method two, stack sampling for metals is conducted in the trial burn while feeding known amounts of the regulated metals. If the metal emissions during the trial burn are in compliance with Table 3, the metal feed rates during the trial burn become the operator's limits.

Method three gives the operator the most flexibility. Using defined health-based maximum allowable ground concentrations of the non-carcinogenic metals and site-specific air dispersion modeling, the operator back-calculates acceptable stack metal emission limits. For the carcinogenic metals, the same procedure is used except that the maximum allowable ground level concentrations are based on a total increased cancer risk of 1 in 100 000. The operator then complies with these site-specific limits through methods one or two.

The new standards also add Cl_2 to the HCl standard. The emission limits for HCl and Cl_2 are established by the same methods as non-carcinogenic metals.

The Toxic Substances Control Act

TSCA was promulgated in 1976 to regulate the commerce of chemical substances in the US. The regulations resulting from this act primarily provide a cradle to grave regulatory authority for chemicals produced, used, or imported into the US. The regulations require testing, record-keeping, reporting, and direct regulation of the production, use, labelling, and disposal of chemical substances. Unlike RCRA, the regulatory authority of TSCA is vested entirely with the federal government, *i.e.* no delegation to the states.

In addition to prohibiting the manufacture of PCBs, TSCA also mandated the thermal destruction of PCBs and PCB-contaminated materials. As a part of accomplishing the destruction of existing PCBs, a national permitting program for PCB incinerators was established.

TSCA sets forth very specific technological standards for PCB incineration. The standards are different for PCB-contaminated liquids and solids. The following requirements apply to the incineration of *liquids* contaminated with PCBs at 50 p.p.m. or greater:

(i) CO, O_2, and the combustion temperature must be continuously monitored and recorded and carbon dioxide must also be monitored and recorded periodically, as specified by the EPA;
(ii) a combustion efficiency of 99.9% shall be maintained. Combustion efficiency (CE) is calculated from the monitored CO and CO_2 levels by the following equation:

$$CE = \frac{[CO_2]}{([CO_2] + [CO])} \times 100 \qquad (2)$$

where $[CO_2]$ is stack CO_2 concentration and $[CO]$ is stack CO concentration;
(iii) the rate and quantity of PCBs fed to the incinerator shall be measured and recorded at intervals not to exceed 15 min;
(iv) a 2 s residence time at $\geq 1200\,°C$ and $\geq 3\%$ excess O_2, *or* 1.5 s residence time at $\geq 1600\,°C$ and $\geq 2\%$ excess O_2 in the stack gas;
(v) a system to automatically cut off PCB feed whenever: (1) the temperature or excess O_2 drop below the minimum specified level or (2) the monitoring systems for CO, CO_2, or O_2 fail;

128

(vi) a water scrubber or other approved method for HCl control; and

(vii) a trial burn where stack O_2, CO, CO_2, NO_x, HCl, total chlorinated organics, PCBs, and particulate matter emissions are monitored when the incinerator first burns PCBs.

The requirements for the incineration of *non-liquid* PCBs are: demonstration that the mass air emissions of PCBs are no greater than 0.001 g PCB kg^{-1} of PCB introduced (*i.e.* 99.9999% DRE) and all of the requirements for *liquid* PCB destruction (given above) except the residence time, temperature, and excess oxygen requirements, and the automatic cut-off requirements for temperature and oxygen.

Typically, a TSCA permit will also include requirements that are similar to those given for RCRA hazardous waste incinerators. Most commercial stationary TSCA permitted incinerators in the US also have a RCRA permit.[17]

3 The Public Involvement Process

All of the permitting programs discussed above are required to include the opportunity for public involvement. The biggest challenge facing the US incineration industry today is public acceptance. Public concern and outcry have been a significant barrier to the expansion of the incineration market at a time when demand is escalating. The public participation requirements for the various regulatory programs are similar and the public participation requirements in the RCRA hazardous waste permitting program are provided below as an example.

The first opportunity for public participation in the RCRA hazardous waste incineration facility permitting program is at the submission of the permit application. After an application is submitted, an introductory notice is made to the public. Often, the incinerator operator will hold informal informational meetings with the public at this stage.

The next opportunity for the public comes when a draft permit or an intent to deny a permit is issued. The permitting authority is required to notify the public of the planned action and allow a public comment period of 45 days. If written notice of opposition is received within 45 days, a public hearing must be held and a transcript or tape recording is placed in a public repository. If no written opposition is received, informal public meetings are often held to discuss the permit. Once a final decision has been made, the permitting authority must give a notice of decision to each person that provided written comments. In addition, the authority must issue a response to comments. Once a final permit decision has been made, no public involvement activities are required.

4 Regulatory Impacts

Emissions

The emissions of waste incinerators have been studied intensely in the past fifteen years. Because of the recent research efforts of the EPA, more is known about US

---

[17] US Code of Federal Regulations, Title 40, Part 761.

incineration technology and its environmental performance than virtually any other waste management alternative.[1] Research conducted by the EPA on incineration has shown that a properly designed incinerator is a very effective treatment device that can be operated safely and with negligible environmental impact or health risks.[2] The research and regulatory framework provided by the EPA has also caused a significant improvement in the practice of incineration over the past decade. These improvements include better performance and control, reduced emissions, and standardized sampling and analytical methods.

The new standards for MSW incinerators will result in a significant reduction in emissions from these sources. They are intended to reduce the air emission of certain pollutants by 90% by 1994. They will require scrubbers at *new large* facilities to reduce metal emissions by 99%, organic compounds including dioxins and furans by 99%, acid gases by 90 to 95%, and nitrogen oxides by 40%. Many existing facilities will be required to add scrubbers in order to reduce metal emissions by 97%, organics by 95%, and acid gases by 73%.[18]

Similar regulations for new and existing medical waste incinerators and small MSW incinerators are now under development by the EPA. The CAA requires that these new standards and guidelines address particulate, SO_2, HCl, NO_x, CO, metals, and dioxins and furans. These regulations are scheduled for proposal in mid-1994 and promulgation in mid-1995.[19]

Because of the public's concern about hazardous waste incineration, the stack emissions of these incinerators have probably received the most attention. EPA-sponsored tests and regulatory trial burns have provided detailed characterization of incinerator stack emissions. The emissions receiving the most scrutiny have been organics, particulate, and acid gases because regulatory performance standards are attached to these pollutants. The conclusion derived by the EPA from the significant quantity of emissions data accumulated, is that a well-operated incinerator, boiler, or industrial furnace is capable of achieving the RCRA performance standards.[2] Of the performance standards, the particulate emissions limit has been the most difficult for facilities to attain. In a published compilation of the results of nine EPA tests and fourteen trial burns, eleven of the twenty-three facilities failed to meet the particulate standard.[20]

The attention paid to the organic emissions from hazardous waste incinerators has been focused on the emission of the POHCs and PICs. Because DRE is a measure of the destruction of a particular organic, a 99.99% DRE does not *guarantee* that total organic emissions are controlled to a level that is protective of human health. Organic emissions can result from the incomplete destruction of organics in the incinerator feed and from the formation of compounds from organic fragments remaining in the off-gas. These emissions have been the subject of considerable debate because they cast doubt on the effectiveness of the EPA's current regulatory approach. One EPA study measured and compared the total organic emissions and specific (frequently observed) organic emissions from

[18] E. M. Steverson, *Environ. Sci. Technol.*, 1991, **25**, 1808.
[19] K. R. Durkee and J. A. Eddinger, Proceedings of the 1993 International Incineration Conference, Knoxville, 1993.
[20] US Environmental Protection Agency, 'Permit Writer's Guide to Test Burn Data—Hazardous Waste Incineration', EPA/625/6-86/012, 1986.

Source	PCDD[a]	PCDF[b]
Medical[c]	117–5260	52–30 300
Municipal[d]	1–10 700	2–37 500
Hazardous[e]	ND–16	ND–56
Boiler[f]	ND–1	ND
Lime/cement kilns[g]	ND	ND

[a]Polychlorinated dibenzo-*p*-dioxin.
[b]Polychlorinated dibenzo-*p*-furan.
[c]Ranges from three hospitals.[22]
[d]Ranges from seven MSW incineration facilities.[22]
[e]Ranges from four hazardous incinerators.[3]
[f]Ranges from four boilers operating on hazardous waste.[3]
[g]Ranges from four lime/cement kilns on hazardous waste.[3]
ND—None detected.

hazardous waste incinerators to the same emissions from MSW incinerators, industrial boilers and furnaces, and coal power plants. The data from these tests indicated that for the organic compounds measured there is little inherent difference between waste and fuel combustion or combustion sources.[21]

The organic emission that has received the greatest amount of attention from the public and scientific community over the years is dioxins. Most of the interest has been on MSW incinerators, but hazardous and medical waste incinerators have also been tested. Table 4 presents a comparison of the dioxin and furan emissions of various incineration sources. While the ranges are very wide, it is apparent that the emission levels from medical and MSW incinerators are higher than from hazardous waste burning devices. This is because until recently, MSW and medical incinerators have not been regulated to the degree that hazardous waste combustors have. This illustrates the effectiveness of the EPA's hazardous waste incineration regulatory program with regard to organic emission control.

A dioxin/furan emission limit of 30 ng $(Nm^3)^{-1}$ was recently promulgated for *new* MSW incinerators and a similar standard is expected for medical incinerators in the next two to three years. The limits for *existing* MSW incinerators were set at 60 to 250 ng $(Nm^3)^{-1}$ depending on the size of the combustor and the degree to which fuel is cofired.[15] The US standard is based on total dioxin/furan emissions rather than toxic equivalence. The standard can usually be met by good combustion practices using CO as an indicator and by avoiding the optimum temperature range for formation in the APCE.

The emissions receiving the most EPA attention currently are toxic metals. The particulate standards of emissions regulations have been the means of controlling metals in the past. The EPA conducted field studies on hazardous waste incinerators in the early eighties and found the particulate standard acceptably controlled the risk from metals emissions. Since that time, the EPA has accumulated a considerable amount of waste characterization data that suggest that hazardous waste often contains considerable amounts of toxic

[21] A. R. Trenholm and C. C. Lee, *Nucl. Chem. Waste Manage.*, 1987, **7**, 33.
[22] C. C. Lee, Proceedings of the 1991 Incineration Conference, Knoxville, 1991.

metals. After performing worst-case risk assessments on the metals emission predicted from burning high metal-content waste, the EPA determined that basing metal controls on the particulate standard alone was not protective of human health. They have since proposed risk-based metal emissions limits for hazardous waste incinerators and promulgated them for BIFs burning hazardous waste. Similar limits are scheduled to be imposed on MSW and medical waste incinerators as a result of the 1990 CAA.

Ash. The management of solid residues remains an issue of concern in the incineration industry. The management and disposal of MSW ash is the subject of considerable controversy. MSW incinerator ash often contains enough leachable hazardous metals to qualify it as a characteristic hazardous waste. Disposing of the ash as hazardous waste is considerably more expensive than disposing of it as MSW. Although the EPA's position is that the ash is exempt from hazardous waste regulation, environmental groups have challenged that position in court. The court and appeals process have resulted in conflicting rulings. These conflicting decisions have left MSW incinerator operators across the country confused over how to manage the ash. Now, the Supreme Court has agreed to resolve the dispute. The decision could have a profound effect on the economic feasibility of MSW incineration.

The challenges facing producers of hazardous waste incinerator ash are more technical in nature. The land disposal restriction regulations require that the organic contaminants be destroyed and the metal contaminants immobilized to defined standards before disposal. These standards are defined for each of the EPA's hazardous waste categories. Since many incinerators burn mixtures of EPA-defined waste streams, the resulting ash must meet the most stringent treatment standards of all the waste categories burned. Testing to verify that the residues meet the treatment standards is a challenging, time-consuming, and expensive task. The operators must simultaneously provide a high level of assurance that the standards are being met while controlling analytical costs, residue storage limits and cost, and test turn-around time.[23] In addition, many operators are having to turn to ash vitrification and other expensive ash treatment methods to meet the metals immobilization standards.

Risk. Much attention is focused on the emissions of various pollutants from an incinerator and the ability of the incinerator to comply with the regulatory treatment standard. While these aspects are important, what is most important is the public health risk that is imposed by the emissions. The EPA conducted a risk assessment to examine the health and environmental effects of the 1982 RCRA incinerator regulations. The analysis used the results of emissions data from nine full-scale incinerator tests. Similar analyses have been done on several MSW incinerators. Table 5 presents a summary of the results of incinerator risk assessments. The hazardous waste incinerator data show that the increased cancer risk due to metals may be as much as five orders of magnitude greater than the risk of organics. It is this apparent risk that prompted the EPA

---

[23] W. Schofield, Proceedings of the 1991 Incineration Conference, Knoxville, 1991.

Total excess lifetime
cancer risk due to
incinerator emissions

Incinerator	Lifetime excess cancer risk
Hazardous waste[24]	
Organics	10^{-10} to 10^{-7}
Metals	10^{-8} to 10^{-5}
Total	10^{-8} to 10^{-5}
Municipal solid waste[25]	
Organics	10^{-7} to 10^{-4}
Metals	10^{-9} to 10^{-4}
Total	10^{-7} to 10^{-4}

to propose risk-based metal emissions limits for RCRA regulated incinerators.

The total risk due to MSW incinerators may be higher than hazardous waste incinerators, although it is very difficult to compare risk numbers (see Table 5). A comparison that can be made, however, is the contribution of individual contaminants to total risk. While toxic metals apparently dominate the risk associated with hazardous waste incinerator emissions, the total risk from MSW incinerators appears to be more evenly split between metals and organics. The organic contributing the greatest risk from MSW incinerators is usually dioxin/furans,[25] while the risk due to dioxin/furans in hazardous waste incinerators is very low. It is important to note, however, that these risk assessments were done before the promulgation of organic emissions control regulations to MSW incinerators. The additional risk incinerators pose to the public health will decrease over the next three to five years as compliance with these regulations is achieved. Similarly, the enforcement of the proposed metal emission controls on hazardous waste incinerators and new metal emission controls on MSW and medical incinerators in coming years will affect the risk from these devices. The net effect of these regulatory changes should be a reduction in total measured risk of at least an order of magnitude.

The risk assessments to date show that a well operated, well designed incinerator presents acceptable risk to public health. This is, of course, dependent upon the perception of acceptable risk. The EPA's definition of acceptable risk is an additional lifetime (70 year) individual cancer risk to the potential maximum exposed individual* of 1 in 100 000. The public's perception of acceptable risk usually approaches zero more closely when it comes to incinerators.

The question also arises of whether incinerator risk analyses truly represent the risk. Risk assessments have to rely to a great extent on conservative assumptions and professional judgement because of deficient data and a lack of guidelines for methodology. A small fraction of the necessary chronic low-level exposure effects data is presently available. There is also considerable uncertainty involved in

---

* The potential maximum exposed individual is defined as an individual located at the off-site location where ambient pollutant concentrations created by a facility are highest, even if this location is not currently populated.

[24] E. T. Oppelt, *Proceedings of the 79th Air Pollution Control Association*, Minneapolis, 1986.

[25] A. Levin, D. B. Fratt, A. Leonard, R. J. F. Bruins, and L. Fradkin, *J. Air Waste Manage. Assoc.*, 1991, **41**, 20.

extrapolating the available data of the effects of high doses on animals to low doses on humans. The lack of knowledge on the interactive effects of complex mixtures of pollutants compounds the uncertainty.[2] A risk assessment guidelines document for hazardous waste incineration is currently being written that will at least standardize the general methodology and eliminate some of the uncertainty.[3] With standardized methodology, the risk assessment result is a useful index of risk even if it is not a quantitative measure.[18]

Technology

The new regulations requiring more stringent controls on a wider range of pollutants are driving the US incineration industry to higher technological levels. Perhaps the most universal and costly effect is the need for APCE. Before the new standards imposed by the CAA and BIF regulations, many facilities were able to meet the emission standards without APCE. Now, many facilities will be looking at adding scrubbers to their systems. This is particularly true for MSW incinerators because of the HCl, SO_2, and NO_x standards.

The metal emissions standards have also affected the types of scrubbers employed. The most common type of APCE employed in incineration has historically been wet scrubbers (*e.g.* venturis and packed beds). Now, primarily because of their high toxic metals and HCl removal efficiencies, semi-dry scrubbers (spray dryer absorbers and baghouses) may become the industry standard.

In the case of medical waste incineration, the CAA regulations will revolutionize the industry. Besides air pollution control retrofits, medical waste incinerators will be pushed into the twentieth century with regard to incinerator operations. Many medical incinerators have been run by untrained operators using much less than state-of-the-art controls and procedures. In order to meet the expected performance standards, this will change or many medical waste incinerators will be forced to shut down.

Industry

The rapidly changing regulatory climate is having a tremendous impact on the incineration industry. One of the major impacts is cost. The recently promulgated standards for new and existing MSW incinerators will increase the cost of processing by an estimated 20%. This is the result of the need by many operators to add or replace scrubbing equipment to keep up with the regulations. The increased cost of medical waste incineration will probably drive many hospitals out of the medical waste incineration business in favor of larger commercial incinerators located to serve a wide area.

Another significant impact is the uncertainty operators face with regard to air emissions. The control equipment installed at significant cost to meet the current requirements may have to be upgraded or replaced within 10 years when the CAA residual risk provisions take effect.

Some regulatory programs have also increased the demand for incinerators in the US. The TSCA regulations mandating the thermal destruction of PCBs are a

134

good example of regulation creating demand. The Land Disposal Restrictions of RCRA have also bolstered the industry by mandating incineration for a large number of waste categories.

Recently promulgated MSW landfill disposal regulations may also bolster the MSW incinerator industry. These regulations, which call for specialized liners, leachate collection, and monitoring systems, increase the cost effectiveness of MSW combustion by increasing the cost of landfill disposal.

5 The Future of Incinerator Regulations

The future will see US incinerator standards evolving from technology-based to risk and health-based performance standards. This has already been set in motion by both the CAA and RCRA. The CAA mandates that, where necessary, the technology-based standards will be replaced by standards that take health risks into account beginning in the year 2001. In many cases, this will result in more stringent emissions standards.

In May 1993, the administrator of the EPA announced a combustion strategy or policy for hazardous waste incinerators and BIFs. This strategy calls for an 18 month capacity freeze, new emission standards for dioxin [$30\,\mathrm{ng\,(Nm^3)^{-1}}$], a more stringent particulate standard [$34\,\mathrm{mg\,(Nm^3)^{-1}}$], and risk assessments that include all pathways. The EPA also announced that by 1996, the agency intends to propose new regulations for hazardous waste incinerators and BIFs. The new regulations are expected to increase public participation and base emissions standards on full and site-specific assessment of health risk.

The EPA called for a capacity freeze because they believe that between incinerators and BIFs, adequate hazardous waste treatment capacity exists. If adequate capacity already exists, any additional capacity could reduce the cost of incineration, which may promote incineration over waste minimization. The EPA's apparent strategy is to promote waste minimization by limiting the growth of the primary treatment option.

In the EPA's announcement, it indicated that health and risk assessments will become a more important part of hazardous waste incinerator regulations in the future. Future emission limits may not be defined in the regulation at all. The emissions limits imposed on a facility may be based entirely on site-specific risk assessments.

Environmental Assessment and Incineration

D. O. HARROP

1 Introduction

Environmental Impact Assessment (EIA) refers to the evaluation of the impacts likely to arise from a development which may adversely affect the environment. The EIA process provides decision-makers with an indication of the likely consequences of their actions. If properly employed, the EIA process will allow informed decisions to be made on planning applications for potentially environmentally significant developments. EIA is thus an anticipatory, participatory environmental management tool.[1] For these reasons EIA has been applied to a plethora of development projects of varying scales. One sphere of development which has utilized the benefits of EIA has been waste management and, in particular, incineration.

To ensure that the significant environmental issues are addressed in a waste management development has necessitated the need for an adaptive EIA procedure. The European Community (EC) Directive (85/337)[2] 'on the assessment of the effects of certain public and private projects on the environment' on EIA has been implemented in the UK through a series of regulations, principally the Town and Country Planning (Assessment of Environmental Effects) Regulations 1988,[3] the Environmental Assessment (Scotland) Regulations 1988,[4] and the Planning (Assessment of Environmental Effects) Regulations (Northern Ireland) 1989.[5] Information on the implementation of the regulations has been given in a series of government circulars and guidance notes.[6] The regulations apply to two specified separate lists of projects. Schedule 1 projects require an EIA in every case and for Schedule 2 projects an EIA is required only if the particular development is judged likely to have a significant effect on the environment. Significant is defined according to the following criteria:

[1] C. Wood and C. Jones, 'Monitoring Environmental Assessment and Planning', Department of the Environment Planning Research Programme, HMSO, London, 1991.

[2] European Council Directive 85/337/EC, 'On the assessment of the effects of certain public and private projects on the environment', *Off. J. Eur. Communities*, 1985, 27 June.

[3] Town and Country Planning (Assessment of Environmental Effects) Regulations 1988 (SI No. 1199).

[4] Environmental Assessment (Scotland) Regulations 1988 (SI No. 1221).

[5] Planning (Assessment of Environmental Effects) Regulations (Northern Ireland) 1989 (SR No. 20).

[6] Department of the Environment, 'Environmental Assessment: A Guide to the Procedures', HMSO, London, 1989.

(i) whether the project is of more than local importance, principally in terms of physical scale;

(ii) whether the project is intended for a particular sensitive location (*e.g.* Site of Special Scientific Interest (SSSI)); and

(iii) whether the project is thought likely to give rise to particularly complex or adverse effects.[6]

Under Schedule 1 of the Regulations (Part 1), a waste disposal installation for the incineration or chemical treatment of special waste requires an EIA. Outwith a Schedule 1 development a developer may submit an EIA voluntarily; otherwise it may fall to the local planning authority to decide whether an EIA is required. Inevitably in the case of most incinerator projects there has been a tendency for the planning authority to request an EIA. If the developer is not satisfied with the planning authority's decision they may seek a direction from the Secretary of State.

Should an EIA be undertaken it is important that the assessment process is thorough and impartial in its execution and its findings. Unless sufficient attention is given to these issues the quality of the resulting Environmental Impact Statement (EIS), the document that reports the findings of the EIA, may be detrimental to the objectivity and credibility of the overall assessment process. The EIS submitted with a planning application in the UK becomes a public document. Therefore it is in the interest of both the developer, their consultant, statutory consultees, the general public, and the environment that approved methods of assessment are employed.[7]

Wathern[8] has broadly identified the following sequence of events involved in progressing an EIA: screening; impact identification; impact prediction and measurement; impact description and evaluation; presentation and communication of information to decision makers; and EIA monitoring and auditing.

The structure of the EIA process is dictated primarily by the need to accommodate each of these key issues, although there may be variations in the detailed procedures adopted within a particular country. Each sequential step is broadly described in more detail below. A generalized procedure is summarized in Figure 1.

2 Screening

The initial step in an EIA process is to determine whether the environmental impact of a proposed project warrants an EIA. This initial step is commonly known as screening. In the UK there are three main procedural stages:

(i) application to the planning authority for an opinion on the need for an EIA;

(ii) application to the Secretary of State for a direction where a developer disagrees with the planning authority's opinion; and

(iii) submission of the EIA.

---

[7] D. O. Harrop and R. P. Carpenter, 'Methods for Assessing Air Quality Impact', Proceedings of the 59th Conference of the National Society for Clean Air and Environmental Protection, Bournemouth, 1992.

[8] P. Wathern, 'Environmental Assessment Methods', Proceedings of the 7th International Intensive Training Course on Environmental Assessment and Management, Centre for Environmental Assessment and Planning, Aberdeen University, 1992.

1 Generalized EIA
procedural system

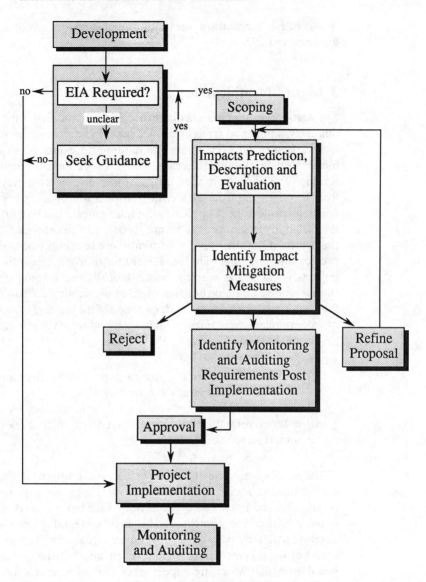

As indicated above, for some developments an assessment is mandatory. Uncertainty arises when it is unclear whether an EIA is required or necessary. A number of countries (for instance Canada, the Netherlands, and Thailand) have incorporated into their procedures the opportunity to assess developments in greater detail in order to establish an EIA requirement. This procedure is referred to as the Initial Environmental Evaluation (IEE) or Initial Environmental Assessment (IEA). The purpose of the IEE or IEA is to supply sufficient information to resolve this basic question. The principal value is that it is a cost-effective mechanism preventing the carrying out of an extensive and unwarranted EIA.

3 Impact Identification

For many practitioners encountering EIA for the first time this may seem a relatively simple and straightforward task. In practice, however, there is often a lack of knowledge concerning the nature and extent of impacts arising from incinerator developments because of their location in different environmental settings. The impacts from an incinerator in one location may be quite different from those arising from an identical installation in another environment. As a result, identification of impacts is complex and should be continued throughout the EIA study as more data becomes available to the project.[9] Nevertheless, from the project outset there is a need to attempt to identify potential impacts at the project initiation stage. The initial identification of the most likely and important impacts is called the scoping stage. Basically the scoping stage of the project should involve dialogue between those implementing the EIA, those responsible for the design, construction, and operation of the incinerator, and representatives of government departments and agencies which have an interest in the development. The objective of the exercise is to:

(i) canvass as wide a body of opinion as practicable, to ensure a comprehensive coverage of local issues and/or concerns;
(ii) focus the study on key issues relevant to the locality; and
(iii) collect information held by certain bodies which is relevant and helpful in undertaking the EIA.

The scoping exercise should provide a description of the most important environmental, social, and economic issues together with the concerns of the community and interested organizations alike in order to describe the potential impacts within the context of the environmental setting of an incinerator development. Not only is this information valuable in gauging whether the likely effects of the scheme will be significant or not, but the exercise identifies those possible impacts which are important enough to deserve a further and thorough assessment. This procedural step is also cost-effective in directing often limited EIA project and staff resources to those areas needing the greatest attention.

There are a range of EIA methods to aid the individual in the identification of potential impacts. These methods are structured mechanisms for the identification, collection, and organization of environmental data.[9] Figure 2 identifies some available methods. Discussed below are those methods that are more easily applied to an incinerator type of development. These and other methods are fully detailed elsewhere.[8,9]

[9] R. Bisset, 'Methods for Environmental Impact Assessment: A Selective Survey with Case Studies', Cobham Resource Consultants, Edinburgh, Proceedings of the 6th International Intensive Training Course of Environmental Assessment and Management, Centre for Environmental Management and Planning, Aberdeen University, 1991.

ure 2 EIA methods
able for identifying
impacts

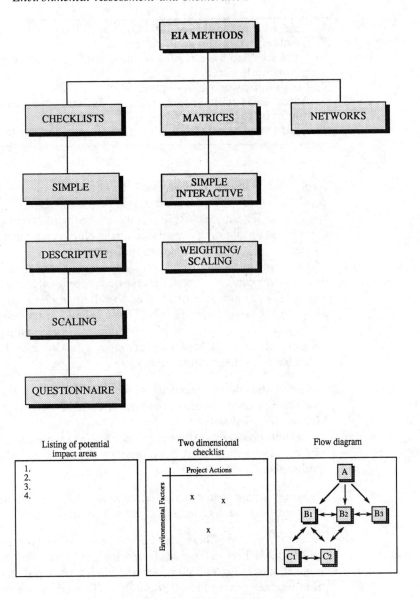

Checklists

Checklists have been widely developed for use in EIA. They are still in general use by many people, either informally or formally, although in many different forms. Essentially they are an initial review of the pertinent aspects of a development which need to be considered. Checklist methods range from listings of environmental factors to highly structured approaches involving importance weighting for factors and the application of scaling techniques for the impacts of each alternative on each factor. Simple checklists represent lists of environmental

Table 1 Waste treatment
and disposal development
checklist

| (1) Lead to considerable pollution of air, soil, or water. |
| (2) Create a risk of spread of disease. |
| (3) Lead to introduction of unwanted wildlife which displaces the original wildlife in the area. |
| (4) Loss of areas worthy of conservation. |
| (5) Affect areas with historic remains or landscape elements which are of importance. |
| (6) Lead to major conflicts with regard to existing land use. |

factors which should be addressed; however, no information is provided on specific data needs, methods for measurement, or impact prediction or assessment.[10] An example of a simple checklist for waste treatment and disposal developments is that based on an example provided by the Norwegian Agency for Development Cooperation (NORAD)[11], shown in Table 1.

In some instances an EIA project manager may have an incomplete knowledge about a development or about the type of environment likely to be affected by the proposed development and will need to call in specialist technical assistance to advise on the assessment. The checklist will help formalize concerns relating to the project. They provide an *aide memoire* to project managers.

Bisset[9] identifies at least four principal types of checklist, each one being applicable to an incinerator development.

Simple lists—these checklists, which contain only a list of environmental factors, are used to focus attention of those undertaking the EIA on those factors (*e.g.* Table 1).

Descriptive checklist—these checklists give guidance on assessment. For each factor, information is provided on appropriate measurements and predictive techniques.

Neither simple nor descriptive checklists offer guidance on how impact importance can be determined. One such technique is a scaling checklist.

Scaling checklist—these checklists consist of environmental elements or resources such as, in the case of an incinerator, air quality, water quality, visual impact, and fauna and flora, accompanied by criteria which expresses values for these resources which are desirable. The method incorporates 'Thresholds of Concern'. The 'Threshold of Concern' value for each environmental element is the point at which those assessing a proposal become concerned with the impact of a particular activity. Any impact which exceeds a 'Threshold of Concern' may be considered to be significant to the decision makers. The 'Threshold of Concern' can represent an objective to be achieved or a limit not to be exceeded. With respect to an incinerator development an appropriate

[10] L.W. Canter, 'Advanced Environmental Assessment Methods', Proceedings of the 13th International Seminar on Environmental Assessment and Management, Centre for Environmental Management and Planning, Aberdeen University, 1992.

[11] Norwegian Agency for Development Cooperation (NORAD), 'Environmental Impact Assessment (EIA) of Development Aid Projects—Checklists for Initial Screening Projects', Oslo, December 1989.

'Threshold of Concern' might be based upon the EC Air Quality directives for sulfur dioxide (SO_2), lead, and nitrogen dioxide (NO_2).

Questionnaire checklists—these checklists consist of environmental elements against which questions on the level of potential impact are asked. There can be one of three answers—yes, no, or if insufficient evidence were available for a definite response, then an unknown category would be marked. If an answer can be determined, then a questionnaire checklist provides a classification for describing estimated impacts: for example, an impact is high or low.

Checklists represent a collective professional knowledge and judgement of the developers of such lists; hence they have a certain level of professional credibility and usability; they provide a structured approach to identifying key impacts and pertinent environmental factors for consideration in EIA; they can be used to stimulate and facilitate inter-disciplinary team discussions during the planning and execution of EIA; and they can be modified to make them more pertinent for particular project types in given locations.[10] Checklists, however, can introduce the danger of creating 'tunnel vision' by only considering the items on the checklist and complex lists have to pay the price associated with technical sophistication. Their use is restricted to those who are familiar with their organization principles, and non-experts may find it difficult to understand and question the results obtained from them.[9]

The principal drawback of checklists is that they deal only with the environment. Attention is focused on only the side of the impact phenomenon. An impact on an environmental component must be caused by a feature or activity associated with the project. These checklist methods do not give any information or guidance on the ways an environmental feature may be affected by one or more development activities. This gap in the coverage of checklists has led to the development and usage of the interactive matrix. The most famous of which is the Leopold matrix.[12]

Matrices

A matrix refers to a display of project actions and activities on one axis with appropriate environmental factors listed along the other. The matrix is, in effect, the embodiment of the fundamental concept of impact prediction, namely that impacts result from the imposition of a project upon a particular area and accrue from the interaction of development activities with components of the local environment.[8] Matrices are in effect two-dimensional checklists. A simple matrix for an incinerator development is shown on Figure 3. When a given action or activity is anticipated to cause change in an environmental factor, this is noted at the intersection point in the matrix. If necessary the impact may further be described in terms of magnitude and importance. The matrix developed by Leopold[12] involved the use of a grid with 100 specified actions and 88 environmental items. For most incinerator developments the simple matrix or a variation of it given in Figure 3 is probably more manageable.

[12] L. Leopold *et al.*, 'A Procedure for Evaluating Environmental Impact', Circular 645, US Geological Survey, Washington, DC, 1971.

Figure 3 Example of an interactive matrix for an incinerator project

Legend:
- ○ no effect
- ● potential adverse effect
- ▦ potential positive effect
- ⊖ effect unknown

Development Activity columns:

SITE PREPARATION AND CONSTRUCTION
1. Traffic generation
2. Construction of access road
3. Site clearing/excavation
4. Drilling and piling
5. Design of plant
6. Installation of utilities
7. Drainage alteration
8. Laying of pipelines for pumping of sludge

COMMISSIONING, MAINTENANCE AND OPERATION
1. Stack emissions
2. Liquid emissions
3. Equipment operation
4. Operational failure
5. Storage of ash

DISPOSAL OF WASTE
1. Disposal of ash to landfill

DECOMMISSIONING
1. Demolition activities
2. Reclamation

EFFECTS ON:

PHYSICAL / CHEMICAL EFFECTS

WATER
A Groundwater
- Flow & Water Table Alteration
- Interaction with Surface Drainage
- Water Quality Changes
B Surface Water
- Drainage Characteristics
- Flow Variation
- Water Quality Changes

LAND
- Soil Quality
- Soil Structure
- Compatability of Land Uses
- Compaction & Settling
- Stability
- Landscape Character
- Geological Resources

ATMOSPHERE
- Air Characteristics
- Wind
- Micro-climate Changes
- Macro-climate Changes

ECOLOGICAL EFFECTS

ECOLOGY
A Terrestrial
- Flora
- Fauna
B Aquatic
- Flora
- Fauna

HUMAN EFFECTS

HUMANS
A Nuisance
- Noise/ Vibration
- Litter / Debris / Dust
- Odour
- Pests / Vermin
B Visual/Recreational Amenity
- Landscape Modification
- Visual Obtrusion
- New Landscape Feature
C Health and Safety
- Health
- Safety
D Socio-economics
- Social Welfare
- Economic Welfare (jobs)

TRANSPORT EFFECTS

TRANSPORT
- Road Capacity
- Road Safety
- Highway Infrastructure

CULTURAL HERITAGE EFFECTS

CULTURAL HERITAGE
- Sites of Archaeological Interest
- Ancient Monuments
- Listed Buildings

In Figure 3 the development actions have been broadly divided up into construction, commissioning and operation, and decommissioning. Identified potential interactions may then be subject to more detailed scrutiny in a more elaborate matrix. Each of the potential interactions are considered in order to determine whether there is likely to be a significant impact. This may involve consultation with experts, the initiation of a preliminary data collection scheme, and discussions with the proponent and local people. In the light of these deliberations, the consequences of each potential interaction can be assessed. The potential interaction may result in either no impact (no effect); an impact which is known and considered not to be significant or can be mitigated (potential adverse effect); a known and significant impact which cannot be mitigated (potential positive effect); or an unknown consequence.

The following general steps should be employed in developing a simple matrix.[10]

(1) List all anticipated project actions and group them in their project timescales:
  (a) construction;
  (b) commissioning;
  (c) operation; and
  (d) decommissioning.
(2) List pertinent environmental factors from the proposed development's environmental setting and group them according to physical, chemical, biological, socio-economic, cultural aspects, *etc.*
(3) Discuss preliminary matrix.
(4) Decide an impact rating scheme (numbers, letters, colours, *etc.*).
(5) Annotate matrix and make notes in order to identify impacts and summarize impacts.

Networks

Of the remaining methods given in Figure 2, networks may be utilized in limited circumstances. Networks attempt to integrate impact causes and consequences through identifying inter-relationships between causal actions and the impacted environmental factors, including those representing secondary and tertiary effects. Figure 4 shows a simple network method applied to an incinerator development. Network analysis is particularly useful for identifying anticipated impacts associated with potential projects. Networks can also aid in organizing the discussion of anticipated project impacts. They are also useful in communicating information to interested people. The primary limitation of the network approach is the minimal information provided on the technical aspects of impact prediction and the means of comparatively evaluating the impacts of alternatives. They are also visually complicated.

4 Impact Prediction and Measurement

The Royal Commission on Environmental Pollution—Incineration of Waste[13]—identifies the principal environmental impacts emanating from an

[13] Royal Commission on Environmental Pollution, Incineration of Waste, Seventeenth Report, May 1993, HMSO, London.

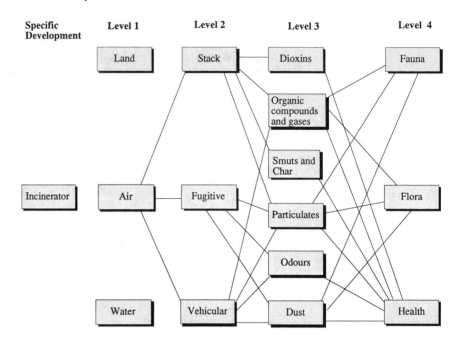

Figure 4 A partial network system for an incinerator development

incineration development as including emissions to air and their associated health implications, visual intrusion, nuisance (odour, noise and smuts, and char), vehicular movements, effects on flora and fauna, and socio-economic effects. Depending upon the environmental sensitivity of the development site, water impacts may also necessitate consideration. Ultimately the environmental concerns will depend upon the proposed environmental setting. The Department of the Environment EIA procedures guide[6] and the statutory regulations identify other issues that may need consideration. They identify the 'specified information' needs for an EIS which include: a description of the proposed development, comprising information about the size and the design and size or scale of the development; data necessary to identify and assess the main effects which the development is likely to have on the environment; and a description of the likely significance, direct and indirect, on the environment of the development with reference to its impact on human beings, flora and fauna, soil, water, air, climate, the landscape, the interaction of any of the foregoing, material assets, and cultural heritage. Where significant adverse effects are identifed with respect to these environmental criteria a description of the mitigation measures to avoid, reduce or remedy the impact should be provided. The need to ensure that each environmental criterion is properly assessed has resulted in the use of disparate scientific techniques to quantify or qualify the level of impact.

For the most part stack gas emissions and their effects on human health in incinerator developments have received and occupied the greatest attention to date in submitted EISs. Therefore due consideration is given in this text to assessing the impacts on ambient air quality.

Air Quality Impact Assessment (AQIA) involves not only identification,

146

prediction, and evaluation of critical variables such as source emissions and meteorological conditions, but also potential changes of air quality as a result of emissions from a proposed incinerator project. It can be used as a screening device for setting priorities for pollution control, or it can be implemented to test alternative project design at an early stage in the planning design and to aid the identification of the most suitable site in terms of mitigating environmental impacts. Through its application it will identify and quantify environmental impacts, and through plant design and planning mitigate for them to ensure that a development's impact on a locality is acceptable.[7]

An AQIA may broadly be divided into three stages as described below. Figure 5 shows a simplified schematic framework based on Stages 1–3.

Stage 1—The Existing Situation

The AQIA assessment begins with a knowledge of the existing situation. This will depend upon: ambient air pollution concentrations; pollutant sources and their specific location; meteorology; local topography; physical conditions affecting pollutant dispersion; and sensitive receptors and their specific location.

In essence, it is important to know what air pollutants are present in the area under consideration and in what quantities, where the pollutants came from, how they will be dispersed, and where they are destined. These information requirements are fundamental to the study.

Ambient Air Pollutant Concentrations.   Concerns arise when air quality data sets are needed for a region not covered by existing or well established monitoring stations. Unless air quality data are available from an unofficial source, then on-site monitoring will be required. This can be both costly and time-consuming and, at best, limited in the data that it is able to offer. Short-term, on-site monitoring may not be representative of seasonal or annual pollution trends. If no data are available then it is prudent to undertake a baseline monitoring study.

Baseline monitoring is the repeated measurement of parameters considered to be important and likely to be affected by the project development. Indeed monitoring must be regarded as a continual integral activity, paralleling project development. The function and subsequent design of effects monitoring programmes must be recognised at the outset of baseline activities and planned appropriately.[14] Baseline monitoring should be planned and initiated during the scoping exercise of an EIA. Monitoring can be integrated with impact prediction and assessment and readjustment carried out as necessary in order to focus on key impacts as the project proceeds.

A basic problem in designing a baseline monitoring programme is that answers to a number of different questions are demanded. For example, the number and location of sampling sites, the duration of the survey and the time resolution of sampling will vary according to the use to which the collected data are to be put. Decisions on what to monitor, when and where to monitor, and how to monitor

[14] M. H. Davies and B. Sadler, 'Post Project Analysis and the Improvement of Guidelines for Environmental Monitoring and Auditing', Report EPS 6/FA/1, Environmental Assessment Division, Conservation and Protection Department, Canada, 1990.

Figure 5 Schematic
framework for an air
quality impact assess-
ment for an incinerator

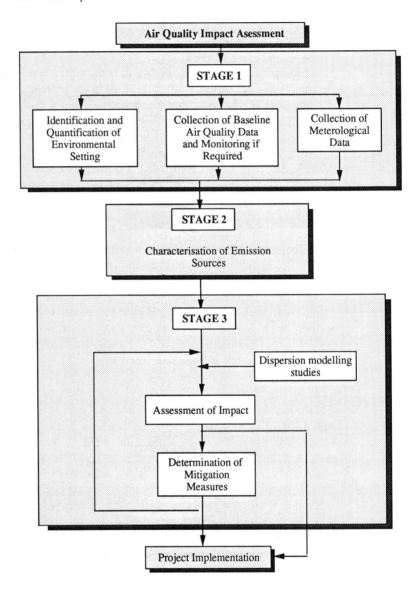

Figure 5 Schematic framework for an air quality impact assessment for an incinerator

must be clearly defined. Therefore it is most important that the first step in the design of a monitoring programme should be to set out the objectives of the study. Once this has been done then the programme may be designed by consideration of a number of steps in a systematic way such that the generated data are suitable for the intended use. It is important also that the data produced by a monitoring programme should be continuously appraised in the light of the study's initial objectives. In this way, limitations in the design, organization, or execution of the survey may be identified at an early stage and where possible remedied.[15]

[15] R. Bisset, 'Role of Monitoring and Auditing in EIA', Proceedings of the 1st Annual Portuguese Seminar on EIA, Albufeira, Centro de Estudos de Planeamento a Gestao do Ambient, April 1991.

The types of ambient air pollutants requiring measurement for an incinerator development include at least those requiring emission control under the Environmental Protection Act 1990 (EPA) and European Community legislation:

oxides of nitrogen (NO_x) (*i.e.* nitric oxide (NO) and nitrogen dioxide (NO_2));
sulfur dioxide (SO_2);
total suspended particulates (TSP);
hydrogen chloride (HCl);
hydrogen fluoride (HF);
carbon monoxide (CO);
volatile organic compounds (VOCs);
dioxins and furans; and
metals: arsenic (As), lead (Pb), copper (Cu), chromium (Cr), nickel (Ni), thallium (Tl), antimony (Sb), cobalt (Co), mercury (Hg), zinc (Zn), tin (Sn), vanadium (V), cadmium (Cd), and manganese (Mn).

It should be remembered that ideally site-specific ambient air quality data are always the most preferable in any AQIA study although they are frequently the most difficult to obtain. The cost limitations frequently imposed upon the project frequently determine the extent of baseline monitoring.

Pollutant Sources and Their Specific Locations.   In many locations, knowledge of principal pollutant sources is often available. Unfortunately, the precise nature of these emissions is frequently difficult to obtain. The near total absence of regional and local pollution emission inventories further hinders the information gathering process. Nevertheless, if an emission inventory exists for adjacent sources in the locality it will further help to establish the impact of the incinerator.

Meteorology.   Whether the proposed AQIA predictive modelling method used is short-term or long-term determines the meteorological requirements for the study. Site-specific data are often non-existent and therefore on-site monitoring is usually the most suitable remedy to the problem. This is frequently costly and therefore may be restricted in the parameters monitored.

The most common meteorological parameters used in air dispersion modelling studies are: windspeed; wind direction; mixing heights; atmospheric stability coefficients (Pasquill stability frequency analysis); and ambient air temperature.

The provision of these data in sufficiently large data sets is often restricted to the Meteorological Office monitoring sites. Unfortunately these data may not be wholly suited to the development site.

Local Topography.   Maps provide most topographical data for study purposes. However, concerns for topographical influences on air dispersion arise when the air dispersion model used may not be sophisticated enough to accommodate a scenario of diverse terrain.

Physical Conditions Affecting Pollutants Dispersion.   Micro- and macro-meteorological data (*e.g.* temperature inversions) may not be available for the development site. The idiosyncrasies of topography affecting local meteorology and climate are often beyond the resources of most studies to identify, although where possible they should be assessed.

Sensitive Receptors and Their Specific Location. Details of the environmental sensitivity of possible receptors are needed to evaluate the impacts of pollutant source emissions. Particular reference to terrestrial and aquatic ecosystems and their conservation value are essential. Sensitive receptors may also include residential areas; schools and recreational/leisure areas; and notified sites of environmental interest (SSSI); *etc.*

Stage 2—Characterization of Emission Sources

The next step of an AQIA is to determine the character of incinerator emissions that will be released from the development. This will include nature of pollutants; emission concentration; emission rate; efflux velocity; efflux temperature; and source morphology (stack height, diameter, *etc.*).

Frequently a combustion source will emit pollutants at varying concentrations, particularly if the feedstock is heterogeneous. Where possible source emission data should reflect actual operations. Therefore the data should represent a range of operating conditions, including those leading to maximum emissions.

Stage 3—Assessment of Impacts

This stage is the assessment of air quality impacts resulting from the identified source emissions from the proposed development. The assessment is generally based on a comparison of the ambient air quality standards (AAQS) for the pollutant of concern to the cumulative concentrations (background and predicted incremental concentrations) of that pollutant. In order to avoid exceeding the AAQS, mitigation measures should be incorporated into the project at the design stage.

An effective method in AQIA procedures is atmospheric dispersion modelling. Many air pollutant problems are often best solved by monitoring. However, this is expensive in terms of staff time, equipment, and analytical laboratory costs. One relatively inexpensive and increasingly used alternative is to employ computer models which can simulate the dispersion of air pollution into the atmosphere.[16] The objective of modelling is to relate mathematically the effect of source emissions on ground-level pollutant concentrations, and to establish whether permissible levels are, or are not, being exceeded once background concentrations are accounted for.

Models have been developed to meet these objectives for a variety of pollutants, time scales, and operational scenarios. Short-term models are used to calculate concentrations of pollutants over a few hours or days, and can be employed to predict worst-case conditions (*i.e.* high pollution episodes). Long-term models are designed to predict seasonal or annual average concentrations, which may prove useful in studying health effects, impacts on

---

[16] D. O. Harrop, 'Tackling Air Pollution Problems with Computer Models', *London Environ. Bull.*, 1986, **3(4)**, 11–12.

vegetation, and materials and structures.

The problem of predicting the distribution of airborne material released from a source is commonly approached by solving the diffusion/transport equation. There is a range of models which have been developed to solve the equation, depending on simplifying assumptions made and the boundary conditions imposed. One type of model widely used is the Gaussian, where the spread of a plume in the vertical and horizontal direction is assumed to occur by simple diffusion perpendicular to the direction of the mean wind.[17] The Equation (1) is the most common expression given. The concentration, χ, of gas or aerosol (particulate matter less than $20\,\mu$m in diameter) at location X, Y, Z from a continuous source with an effective emission height, H, is given by:

$$\chi(X,Y,Z;H) = \frac{Q}{2\pi\sigma_y\sigma_z u}\exp[-1/2(Y/\sigma_y)^2]\{\exp[-1/2((Z-H)/\sigma_z)^2] + \tag{1}$$
$$\exp[-1/2((Z+H)/\sigma_z)^2]\}$$

H is the height of the plume centre line when it becomes essentially level, and is the sum of the physical stack height and the plume rise. The following assumptions are made of the Gaussian model: the plume spread has a Gaussian distribution in both the horizontal and vertical planes, with standard deviations of plume concentration distribution in the horizontal and vertical of σ_y and σ_z, respectively; the mean windspeed affecting the plume is u; the uniform emission rate of pollutant is Q; and total reflection of the plume takes place at the earth's surface (*i.e.* there is no deposition or reaction at the surface).[17]

For concentrations calculated at ground level (*i.e.* $Z = 0$) and along the centreline of the plume (*i.e.* $Y = 0$) Equation (1) simplifies to:

$$\chi(X,0,0;H) = \frac{Q}{\pi\sigma_y\sigma_z u}\exp[-1/2(H/\sigma_z)^2] \tag{2}$$

Using equation (2), it is possible to carry out a simple and rapid screening assessment of a proposed development.

5 Impact Description and Evaluation

The primary objectives of the study are to protect public health, flora and fauna, property, and amenity. The need to protect health arises out of the observed acute ill effects of pollution episodes and the results of epidemiological studies into the long-term effects of chronic exposures to relatively low concentrations of pollutants.

Within the UK statutory ambient air quality standards are limited to the EC Directives on nitrogen dioxide (EC Directive 85/203), lead in air (EC Directive 82/884), and sulfur dioxide and smoke (EC Directive 80/779). Outwith these standards, air quality guidelines are used to assess the effects of air pollutants on health, and the World Health Organization (WHO) guidelines generally receive the greatest attention. A recent publication has summarized the AAQS for a

[17] D. B. Turner, 'Workbook of Atmospheric Dispersion Estimates', Air Resources Field Research Office, Environmental Science Services Administration, Environmental Protection Agency, Offices of Air Programs, Research Triangle Park, North Carolina, 1970.

number of countries.[18] A widely used 'rule of thumb' is the employment of the relevant UK Health and Safety Executive's (HSE) EH40 Occupational Exposure Limits (OEL) divided by a factor of either 40 or 100, depending upon the quality of air in the study area. In this manner, the incorporation of a safety factor allows workplace standards to be extrapolated to the external environment. Where possible, other national and international AAQS are also employed.

The threats to fauna and flora are similarly of concern, particularly with complex environmental interactions such as the discharges of acid gases from incinerators, the transformation of these gases to form acid rain, and the indirect impacts on sensitive aquatic and solid chemistry systems due to the precipitation of such rain in environmentally sensitive areas.

The possible injury and damage to plant communities is a combination of a range of physical, chemical, and biological stresses which may affect the plant's physiology. The visible symptoms produced by these various stresses need to be distinguished, as do the very different symptoms which can be produced in different plant species by the same factor. Further difficulties are encountered by the fact that plants are commonly subjected to more than one stress, either simultaneously and successively, and that the sensitivity of plants to a particular stress will be altered by other environmental factors. When assessing the impact of air pollution no two incidents in the field are the same; each involves a unique combination of the concentration and the exposure period to the pollutants, of plant species, plant age, and environmental conditions.[19] It is therefore difficult to identify a direct cause and effect relationship between an air pollutant concentration and its effect on a plant. Nevertheless, like many of the above-mentioned potentially significant environmental effects of an incinerator development, each must be systematically assessed to quantify the level of impact.

6 Presentation and Communication of Information

Perhaps the most important activity in the EIA process is the presentation of the EIS. The EIS aids the decision-makers in their final decision relative to the particular project, and being a public document it will be scrutinized by interested agencies and groups. Therefore, it is critical that special care is taken in the preparation of the statement. Canter[20] summarizes from other authors five basic principles to be remembered:

(i) always have in mind the audience of the report; in the case of an EIS assume that the reader is intelligent but uninformed;

[18] L. Murley, 'Clean Air Around the World', 2nd Edition, IUAPPA, Brighton, 1991.
[19] H. J. Taylor, M. R. Ashmore, and J. N. B. Bell, 'Air Pollution Injury to Vegetation (A Guidance Manual Commissioned by HM Industrial Air Pollution Inspectorate of the Health and Safety Executive), Institute of Environment Health Officers, London, 1988.
[20] L. W. Canter, 'Preparation of EIA Reports (Technical Writing Principles)', Proceedings of the 2nd International Course on Management of Environmental Conflicts and Impact Assessment, (Module III—Advanced Course on EIA Methods and Technologies), Bologna, Italy, April 1991.

(ii) decide on the purpose of the report, to convey the environmental consequences of the proposed development;

(iii) use simple and familiar language, the EIS requires the submission of a non-technical summary;

(iv) ensure that the presentation of the report is well-structured; and

(v) make the report visually attractive.

The EIS should follow a logical process and be prepared in a consistent manner which can aid in communicating information to both technical and non-technical audiences.

7 EIA Monitoring and Auditing

EIA remains a predictive process, focusing on the identification and assessment of changes in environmental systems resulting from a development proposal. The changes that actually occur at the post-development stage, however, are rarely considered, far less related to the effects anticipated at the outset of the project. Due to these limitations, scientific credibility of the EIA process has been lowered.

There is a realization that the impact information upon which decisions on projects are, in part, made is open to a degree of uncertainty. Therefore an analysis of actual impacts in the post-project phase can do much to enhance the predictive ability and credibility of future EIA studies.[21] It is only with some form of systematic follow-up to decision-making, closely linked with the EIA process, that an effective and meaningful environmental planning and management system can be achieved. The objective is therefore to instigate an approach whereby appropriate information derived from impact monitoring studies is fed back into the EIA process to achieve improvements in the identification and assessment of project-induced impacts. Such arrangements provide continuity in the process, linking the pre-decision and post-decision phases of the project cycle. Monitoring and auditing, therefore, not only serve to improve the management of projects by refining mitigation measures, but also facilitate learning from experience, and improve the process and practice of EIA as a management exercise.[22]

Selected Reading

1 R. G. H. Turnbull (ed.), 'Environmental and Health Impact Assessment of Development Projects—A Handbook for Practitioners', Elsevier Applied Science, London, 1992.

2 P. Wathern (ed.), 'Environmental Impact Assessment—Theory and Practice', Unwin Hyman, London, 1988.

3 J. Petts and G. Eduljee, 'Environmental Impact Assessment for Waste Treatment and Disposal Facilities', John Wiley and Sons, Chichester, 1994.

[21] R. Bisset, 'Monitoring and Auditing of Impacts: A Review', Cobham Resource Consultants, Edinburgh, Proceedings of the 6th Intensive Training Course on Environmental Assessment and Management, Centre for Environmental Management and Planning, Aberdeen University, 1991.

[22] R. G. H. Turnbull, 'Environmental Assessment: Monitoring and Auditing', Paper Presented at Environmental Assessment Course, Technical University of Budapest, Budapest, Hungary, February 1993.

Subject Index

Subject Index